The Best American Science and Nature Writing 2008

The Best American Science and Nature Writing™ 2008

Edited and with an Introduction
by Jerome Groopman

Tim Folger, Series Editor

HOUGHTON MIFFLIN COMPANY
BOSTON · NEW YORK   2008

www.hmhbooks.com

ISSN 1530–1508
ISBN 978–0–618–83446–4
ISBN 978–0–618–83447–1 (pbk.)

Printed in the United States of America

MP 10 9 8 7 6 5 4 3 2 1

Jon Cohen, "Zonkeys Are Pretty Much My Favorite Animal." First published in *Outside,* August 2007. Copyright © 2007 by Jon Cohen. Reprinted by permission of the author.

John Colapinto, "The Interpreter." First published in *The New Yorker,* April 16, 2007. Copyright 2007 by John Colapinto. Reprinted by permission of the author.

Christopher J. Conselice, "The Universe's Invisible Hand" from *Scientific American,* February 2007. Reprinted with permission. Copyright © 2007 by *Scientific American,* Inc. All rights reserved.

Gareth Cook, "Untangling the Mystery of the Inca" from *Wired,* January 2007. Copyright © 2007 by Condé Nast Publications. Reprinted by permission of Gareth Cook.

C. Josh Donlan, "Restoring America's Big, Wild Animals" from *Scientific American,* June 2007. Reprinted with permission. Copyright © 2007 by *Scientific American,* Inc. All rights reserved.

Freeman J. Dyson, *A Many-Colored Glass: Reflections on the Place of Life in the Universe,* pp. 7–9, 13–21. © 2007. University of Virginia Press. Reprinted with permission. As published in *The New York Review of Books,* July 2007, as "Our Biotech Future."

Steve Featherstone, "The Coming Robot Army." Copyright © 2007 by *Harper's*

Contents

Foreword

ON THE EVE of the solstice two summers ago, I saw something that in ages past would have evoked both wonder and fear. For three minutes, two points of light moved like errant stars eastward across the night sky just north of the Big Dipper. Centuries ago this would have been an omen of the most ominous sort, auguring the fall of a kingdom, perhaps, or something even direr. But the spectacle of two objects coursing against the ceaseless westward flow of everything else in the heavens drew no crowds into the streets of my dark and quiet neighborhood. The lights vanished below the horizon at 10:27 P.M. Mountain Time with no wailing, beating of drums, or alarms of any kind. My wife and I were the only witnesses on our street as the space shuttle *Atlantis* and the International Space Station hurtled along at 17,000 miles per hour some 210 miles above northern New Mexico, easily visible to the naked eye.

Earlier that evening the shuttle had separated from the space station after ten days in orbit. News of the space walks, the installation of an enormous array of solar panels, and the mission's other tasks were buried in the back pages of papers or as links on Web sites, if covered at all. The front page of the next morning's *New York Times,* for example, included stories about Tony Blair, New Orleans, and violence in Iraq, but not a word about several humans orbiting the planet every ninety minutes or so. No videos of the crews' views of Earth, no shots of the shuttle pulling away from a football-field-size space station. The dearth of coverage reminded me of something the late Richard Feynman wrote forty-four years ago in an introductory physics text, *The Feynman Lectures on Physics:* "From a long view of the history of mankind — seen from, say, ten thousand years from now," he wrote, "there can be little doubt that the most significant event of the nineteenth century will be judged as Maxwell's discovery of the laws of electrodynamics. The American Civil War will pale into provincial insignificance in comparison with this important scientific event of the same decade."

I'd side with Feynman in a bet that the story of the fitful beginnings of space exploration — and, more generally, the sorts of developments written about in this anthology — will be remembered long after Brangelina, the downfall of Eliot Spitzer, and other fixations of the moment are forgotten. Christopher Conselice, for example, reports on one of the most important — and strangest — scientific discoveries in decades, namely that most of the universe is composed of something called dark energy. Physicists don't know what it is yet, but they do know it's everywhere — in our kitchens and in intergalactic space, as Conselice tells us in "The Universe's Invisible Hand." Conselice, himself an astronomer, writes that cosmologists now believe that dark energy played a crucial role in the formation of galaxies. Without dark energy, our own galaxy might have been destroyed eons ago. And in the distant future, dark matter will . . . well, I'll let Conselice tell that story.

Stepping back from the cosmic scale, Freeman Dyson makes some provocative predictions about a future in which genetic engineering kits may be as common and widely used as personal computers and cell phones are today, with even children creating new life forms. Will that future, if it comes to pass, become the stuff of dreams or of nightmares?

Another glimpse of the future comes from Steve Featherstone, whose observations of army field tests of advanced weapons systems in "The Coming Robot Army" make for disconcerting reading. Equally unsettling is Robin Marantz Henig's "Our Silver-Coated Future," in which she describes how nanotechnology, a $3 trillion industry, is fast becoming ubiquitous, with barely any discussion of its safety.

The other twenty stories in this year's collection are just as compelling. A movie could be made from James Geary's thrilling and sobering account of the murder of a former Russian spy in "The First Assassination of the Twenty-first Century." Jon Cohen encounters a number of animals I'd never heard of, including zorses, tigons, pizzlies, and other intriguing hybrids in "Zonkeys Are Pretty Much My Favorite Animal."

So there's quite a range here, from the cosmic to the tragic. And

I think all these stories will bear rereading in the years ahead, which cannot be said of most stories in the media. As I write, I'm still haunted by the ending of Christopher Conselice's story about the fate of the universe. But Earth will have vanished long before the universe begins to unravel. Astronomers tell us that 5 billion years from now the sun will swell into a red giant and incinerate our planet. It may be that the only human artifacts to survive that cataclysm — the only evidence that we were ever here at all — will be drifting space probes like *Pioneer 10,* launched in 1972 on a twenty-one-month voyage to Jupiter. The spacecraft is now nearly 9 billion miles from home, somewhere near the edge of the solar system, its signals too weak to be detected. But it's out there, and will be forever, carrying a message from Earth inscribed on a gold and aluminum plaque in case some extraterrestrial intelligence ever stumbles upon it. The odds of that happening? Astronomical. The late Carl Sagan and his wife designed the plaque, which includes a diagram of the solar system and the figures of a man and woman standing next to a drawing of the spacecraft. I can think of no more wildly optimistic and life-affirming gesture than the inclusion of that plaque on *Pioneer 10.* Maybe hope isn't a thing with feathers; maybe it's a thing that has traveled 9 billion miles with a message that may never be read.

I hope that readers, writers, and editors will nominate their favorite articles for next year's anthology at http://timfolger.net/forums. The criteria for submissions, deadlines, and the address to which entries should be sent can be found in the "news and announcements" forum. I also encourage readers to use the forums to leave feedback about the articles in this collection and to discuss all things scientific. The best way for a publication to guarantee that its articles are considered for inclusion in the anthology is to place me on the subscription list, using the address posted in the news and announcements forum.

It has been a pleasure to work with Jerome Groopman. Were he not our guest editor, I'm sure one of his articles would have been included in the anthology. I'm grateful to Amanda Cook and Elizabeth Lee at Houghton Mifflin for helping to put together this wonderful anthology. And I'm looking forward to many more nights of stargazing with my beauteous wife, Anne Nolan.

TIM FOLGER

Introduction

ONE OF MY professors in college said, on the first day of class, "To the educated mind, no two things are unrelated." At the time it sounded like a pompous remark, designed to impress a group of young and still naive students. But I eventually came to realize that my reaction was overly harsh. We, as human beings, live in a connected world. Every part of that world — whether immediate and palpable, like the animals and plants and soil we encounter each day, or more distant, perceived through telescopes and microscopes — may be unexpectedly linked. It takes an inquiring and open mind to appreciate connections that are not readily apparent.

In selecting the articles in this anthology, I sought such surprising and significant connections. Each piece reveals ties that many of us do not ordinarily appreciate. Some of these ties are between elements and forces in chemistry and physics; others are in biology, between human beings and the animal kingdom; still other connections are anthropological and archaeological, between human beings and their cultures, present and past.

Writing about science and nature is not at all easy, as I learned some twelve years ago. I had spent more than two decades as a physician and researcher in the fields of blood diseases, cancer, and HIV. Preparing technical papers for clinical and scientific journals was essentially formulaic; you followed a set structure and filled it with "materials and methods" and the resulting data. The core of an article in the *New England Journal of Medicine* or *Science* or the *Journal of Biological Chemistry* was the tables and graphs, with the data points bracketed by statistics. It was the statistical analysis that gave the experimental results "significance," meaning that very likely the observations were not the result of chance alone but reflected some true aspect of the biological or clinical phenomenon.

After many years of publishing such articles, I had something of a midlife crisis. I was in my midforties, increasingly restless, and looking for a new venue. Unlike others of my age, it was not a sports car or skydiving that appealed to me. I imagined becoming an author and decided I would write not only for the *Journal of Biological Chemistry* but also for a lay audience. My subjects would be found in the clinic and the laboratory, where I had encountered remarkable people who faced difficult and uncertain circumstances, having fallen ill with a potentially fatal disease such as leukemia or AIDS. As they confronted their mortality, these people looked to physicians and scientists like me for answers. It was the search for those answers that I as an author wanted to depict.

So for several months I sat at the kitchen table at first light and banged on my laptop. Fueled by several cups of high-test coffee, I produced three stories about patients and their maladies. In each tale I tried to capture both the character of the person and the biology of their disease. I believed I had succeeded and was ready to present my work to the world.

My first readers were colleagues at the hospital. I showed my three drafts to researchers and doctors and received glowing feedback. "Brilliant," said one; "deeply engaging," said another; "profound," remarked a third. Feeling rather heady, I showed the three pieces to my wife, Pam, who studied biochemistry in college and is now on a university faculty as an endocrinologist. She looked at the articles intently, then fixed her gaze on mine. "They're awful," she said. "Really awful. They're convoluted and filled with jargon. Didn't you ever learn how to write a simple declarative sentence?"

"But the people I showed them to found them brilliant, deeply engaging, profound," I objected.

"Who told you that?"

I gave her the names.

"They all work for you."

Pam was right. The three pieces were awful — verbose, often tangential, filled with scientific lingo that would mean nothing to a lay reader, and, on second reading, opaque. The familiar formulaic structure of a scientific article had given me no clue how to write a piece for the average reader; here significance was not a matter of statistics.

I sat down again at the kitchen table and forced myself to rewrite each piece. I generated scores of drafts, straining to craft the structure and sentences that would tell a story and communicate something "significant." The words did not just flow, and even after all these years, they still don't.

Some of the most helpful advice on science writing came from an editor at *The New Yorker* when I began submitting regularly to that publication. I suspect the editor knew that scientists are schooled in the relationship between structure and function, so he took the time to explain the "three pillars" of a narrative's "architecture" and how they support a solid plane of science writing. First was what he called "the argument," meaning the overarching theme. Second was the "protagonists," meaning the voices that would articulate the theme. And third was "cinema." This proved to be the hardest pillar to erect. "Cinema" meant using words to paint pictures in the mind of the reader, so that she felt she had entered the page and was standing with me at the patient's bedside in a moment of crisis or peering into the microscope when the biopsy revealed a cluster of cancer cells.

The articles in this anthology all stand on these three pillars. They have novel and surprising arguments, protagonists who articulate their themes in clear, cogent voices, and vivid cinema. They are not verbose or tangential. They are filled with simple declarative sentences. In a nod to my college professor, each piece also educates us and invites us to make connections among elements and events that previously seemed disparate or unlinked. And, although I cannot be sure, I suspect that none of the articles was easy to write. Each shows a depth of thought and reporting that takes time and considerable effort. In some instances, the author traveled to the far reaches of the globe to visit animals in their native habitat or gauge the impact of the industrialized world on developing countries or stand at an archaeological dig. Other voyages were those of the imagination, seeking metaphors to explain unseen forces in the cosmos or the nature of submicroscopic particles.

Selecting the diverse articles for this collection proved not to be a burden but rather a joy, because of their richness and depth. It is a challenge to select "the best" from the spectrum of scientific inquiry, from physics and cosmology to zoology, anthropology, and human biology. Naturally, when I received the stack of potential picks from the series editor, Tim Folger, there were too many wonderful pieces, so limits of space and category forced many of my decisions. I began big, designating as the first category the cosmos (nothing is larger). Astrophysics has always filled me with awe. In "The Universe's Invisible Hand," Christopher Conselice describes how dark energy not only accelerates and expands the universe but also sculpts the contours of the galaxies. And he does not simply ponder its distant effects: dark energy, he points out, might evolve and become increasingly dominant over time, disrupting entire galaxies. "Ultimately, planet Earth will be stripped from the sun and shredded . . . Dark energy, once cast in the shadows of matter, will have exacted its final revenge." Similarly, Jon Mooallem, in "A Curious Attraction," takes us on the hunt for a possible force that has eluded and perplexed physicists since Einstein's breakthroughs a century ago: antigravity. Is antigravity the unicorn of physics? Or do we simply lack the strategy and bait to catch it in an experimental snare?

Exotic cosmic forces aren't the only ones that may affect the future of our planet. Familiar human endeavors found in our carbon footprint will as well. The Nobel Committee this year recognized the crisis of global warming, and fierce debates rage between advocates and skeptics as the world looks for palatable political and economic solutions. Sometimes the local may inform the global, as Florence Williams shows in "A Mighty Wind." She profiles the small Danish island of Samsø and convincingly argues that it is possible to be clean, green, and energy-independent by melding new technology with an entrepreneurial spirit as we work to reclaim the environment. "Some islands feel magical," Williams writes, "because of healing waters or succulent fruits. Samsø's aura comes from the fetching way the islanders crunch their numbers," in the attempt to offset the carbon dioxide they add to the atmosphere. Some of the estimates center on the number of trees they need to plant; other calculations involve their production of clean energy in lieu of greenhouse gases.

At each step along the path of scientific progress, there are opportunities for great benefit but also for the abuse and perversion of the discovery. I entered science in the early 1970s, when activists referred to fictional plagues like the Andromeda Strain and fictional monsters like Frankenstein in their clash with molecular biologists who were beginning to harness DNA technology. The Princeton physicist Freeman Dyson, in "Our Biotech Future," emphasizes how important freedom and transparency are in making science both familiar and welcome in modern society, employing as a "proof text" the battles that raged over genetic engineering and, now, stem cell research. Dyson draws a wide circle and brings us, at the end of the piece, to "green technology" and how controversial science can work to benefit the environment. On the other hand, some fictionalized nightmares have become reality — specifically radioactivity. In "The First Assassination of the Twenty-first Century," James Geary sets out in chilling detail the murder by polonium-210 poisoning of a former Russian spy who had emigrated to England. "Nuclear physicists call polonium the Terminator," Geary writes, "not because of its efficacy as a poison but because it's the final element created in the process known as slow neutron capture." Yet as this murder case shows, polonium introduced into the human body can indeed terminate life. Of course, we are best prepared against the misuse of science through open discussion of benefits and risks. Robin Marantz Henig asks hard questions in "Our Silver-Coated Future" about the lucrative nanotechnology industry, focusing on its possible hazards to health and the environment. For example, nanosilver, which has antimicrobial properties, is being used to coat handrails in the Hong Kong subway, baby bottles in Korea, and the keyboards of some computers. No one knows what nanosilver may do, but studies of silver contamination in the past show it can cause sterility in aquatic life and disrupt the food chain. Other nanoparticles are taken up by cells and cause deposition of collagen, the major protein in scar tissue. If someone inhaled the particles, his lungs could become stiff and he would gasp for breath.

Even when technology is not life-threatening or fraught with environmental risks, it can endanger us in lasting ways. "The Autumn of the Multitaskers" is a lively and insightful essay by Walter Kirn about the unintended consequences of devices like computers, iPods, and Blackberries. Kirn marshals powerful evidence, drawing on insights from neuroscience, that the multidirectional bombardment of information, and our response to it by multitasking, impairs creative thinking and makes us ever more frantic — to use the popular term, scatterbrained. (This article struck a particular resonance with me, since I indulge in multitasking as a way to procrastinate.) Constant connection to information and to other people was supposed to provide "efficiency, convenience, and mobility," Kirn points out, but these supposed virtues can conspire to limit freedom rather than expand it. Of course, not all e-mail is well intentioned or from friends or employers. Con artists are increasingly employing the Internet in their scams. Ron Rosenbaum, in "How to Trick an Online Scammer into Carving a Computer out of Wood," provides a delightful and brilliant exposition of the effort to turn the tables on such con artists in what has become a vicious cyberwar. Using computers and other machines to prevail over a foe has also caught the attention of our military planners. In "The Coming Robot Army," Steve Featherstone takes us not into cyberspace but onto a battlefield, where he vividly describes how advanced technology may soon replace soldiers.

Science not only imagines the future but, in the field of archaeology, brings alive the remote past. By animating people and places in history, we can more deeply understand the present and learn how cultures succeed and fail. Andrew Lawler takes us to the time before Constantine, when Christianity was in its infancy; "First Churches of the Jesus Cult" explores Megiddo, the *tel* in northern Israel that gives us the name Armageddon, and other sites where a new faith took form. Contrary to the notion that the early Christians, by breaking with Judaism and deviating from the rulers in Rome, were relentlessly persecuted, archaeological evidence suggests that some Christian communities lived in relative harmony with the surrounding Jews and the Roman conquerors. This is in sharp contrast to the image of "lions dining on martyrs in Rome's Colosseum," Lawler writes.

A very different form of Christianity was practiced by the Spanish conquistadors, who were searching primarily for treasure in this life rather than the afterlife. Gareth Cook shows us a treasure not of gold but of the mind in "Untangling the Mystery of the Inca." Cook points out that the Inca never figured out how to write. The mystery of how they communicated may have been solved by anthropologists and computer analysts studying "strange, once-colorful bundles of knotted strings." These strings, which the Spanish noticed but largely ignored, appear to be a highly sophisticated method of encoding information. Another unique opportunity to expand our understanding of human communication is the subject of John Colapinto's wonderful piece "The Interpreter," which takes us on a voyage to the Pirahã, a hunter-gatherer tribe on a remote tributary of the Amazon. Deconstructing the unique speech of this indigenous people has upended prevailing paradigms about the theory of universal grammar.

The age of exploration moved not only armies of soldiers like the conquistadors but armies of pathogens. Today's battles are increasingly fought against viruses and bacteria resident in lower animals that are then passed to human beings, as David Quammen sharply depicts in "Deadly Contact." "The pathogen may be well adapted to its quiet, secure life within a reservoir host," Quammen writes; "spilling over into a new species presents a chance, at some risk, of vastly increasing its abundance and its geographic reach. The risk is that, by killing the new host too quickly, before getting itself transmitted onward, the pathogen will come to a dead end. But," he continues, "evolutionary theory suggests that some pathogens, on some occasions, will accept that risk in exchange for a big payoff." Although new scourges rivet our attention, Michael Finkel rightly instructs us to sustain focus on a familiar and still devastating mosquito-borne infection in "Malaria: Stopping a Global Killer." The magnitude of good that can be achieved with relatively simple measures like mosquito nets is breathtaking. Sometimes the viruses we acquire through evolution become part of us and work in unexpected ways within our DNA, as Michael Specter shows in "Darwin's Surprise." Here we learn that the parasitic behavior of retroviruses may foster beneficial changes when they enter the human genome. For example, these viral sequences may have fostered the development of the placenta in mammals, so that we have live births. Specter quotes a prominent French virologist, Thierry Heidmann: "It is quite possible that" without retroviruses insinuating themselves into our genome through the course of evolution, "human beings would still be laying eggs." Who would have imagined this connection?

Of course, the animal world offers much more than foreign threats or facilitating nucleotides. It can also act as a mirror of human behavior. Evolutionary biologists study primates in their natural environments and seek to find links to our individual and social behaviors. Ian Parker, in "Swingers," questions whether the widespread characterization of bonobos as the "hippies" of the African apes, allegedly peace-loving, matriarchal, and with insatiable sexual appetites, is actually a projection of our own fantasies. Parker points out how the media love to sensationalize science, especially when it involves sex. On the other hand, what may seem a fantasy can be made into reality by modern technology, as Jon Cohen reveals with wit and humor in "Zonkeys Are Pretty Much My Favorite Animal." Cohen introduces us to hybrid creatures, bred in captivity, including tigons, wholphins, zorses, and camas. But we need not visit such a zoo to be astounded by opportunities in the animal kingdom. In "Restoring America's Big Wild Animals," C. Josh Donlan suggests that we may be able to foster the ecosystem and benefit our society by reintroducing large animals from past millennia to the Americas.

Science is a familiar touchstone in many modern dramas, but how accurate is the depiction of researchers and their methodology? Jeffrey Toobin shines light on this question in "The CSI Effect," arguing that forensic science is often quite different from the glitz of TV drama. The media also frequently exaggerate preliminary research to fill the public's deep desire for "news you can use." This is most clearly the case with epidemiological studies. Woody Allen, in the movie *Sleeper,* wakes up in the future with doctors hovering over him, cigarettes in their hands. That smoking is now established as key in the genesis of many cancers, heart disease, and lung disorders makes the scene in the movie absurd, but the audience laughs because Allen is pinpointing how even the most authoritarian pronouncements are at risk of being reversed. Andreas von Bubnoff rightly illuminates the absurdity of broadcasting premature and often flawed data. He focuses on how coffee was said to increase the risk of pancreatic cancer in 1981 but then didn't in 2001, and how my same favorite beverage was said to reduce the risk of colon cancer in 1998 but not in 2005. His "Numbers Can Lie" is one of the best examinations I have seen of epidemiology and its discontents.

All of us have fantasized that some passion might lead us to fame and fortune. I grew up trying to perfect the grace of a breaking curve ball; others put brush to canvas, and still others worked at hitting high C in the choir. In "A Bolt from the Blue" Oliver Sacks takes us deep into the brain and seeks the origins of sudden intense passions, with his trademark skills as a storyteller and scientist. In one case, a man recovers from being hit by lightning with an insatiable desire to listen to piano music. Sacks writes, "This was completely out of keeping with anything in his past. He had had a few piano lessons as a boy" but had not sustained an interest in playing and did not have a piano in his house. Sacks's genius is to use the rare and bewildering as an entree into issues that are familiar but still unexplained.

Our passions need not center on ourselves. In "Children Are Diamonds," Edward Hoagland explores the passionate altruism of those dedicated to helping poor people in Africa. The origin of such imperatives is certainly a mix of culture and heredity, as Olivia Judson highlights in "The Selfless Gene." Judson sets out to weigh the relevant contributions of nature and nurture in the fabric of a caring community. A culture of altruism is a goal of many religions. While religion and science have been cast as perpetual adversaries in several best-selling books, Todd Pitock looks for détente in "Science and Islam," profiling those Muslims who seek to reconcile domains of empirical knowledge and spiritual meaning for the betterment of humankind.

I read these articles in the early morning, that dark and quiet time before the sun rises, with my cup of freshly brewed coffee (which may or may not be healthy but is certainly enjoyed). Each piece filled me with a sense of excitement; I was like a child opening a present — the gift of learning something entirely new. Or learning that what I thought was true may not be.

Perhaps it was the bracing aroma of the coffee that, in stimulating my olfactory receptors, made me more receptive to these new ideas. We now know that these "smell" circuits are linked to pathways in the brain that integrate logic and emotion, cognition and feeling, the conscious and the subliminal. Perhaps the caffeine widened and accelerated those circuits to enhance the access of new knowledge and to more deeply imprint it. Learning becomes part of us; the information and insights stored in our minds can be retrieved from memory, shared with others, and reflected on, so that ultimately we are able to change the way we understand the world and our place in it.

I'm still not fully convinced that my college professor was right that, in essence, everything is linked. Maybe it is a matter of continuing my education before I can see all the connections. But the articles in this anthology support that view in a profound way. They draw the reader more tightly into the web of the world. They forge links in unexpected ways. They connect us to nature and to each other, and those connections nourish the intellect and uplift the spirit.

JEROME GROOPMAN

JON COHEN

Zonkeys Are Pretty Much My Favorite Animal

FROM *Outside*

A FEW MILES FROM THE ENDLESS MALLS and garish tourist attractions of Myrtle Beach, South Carolina, there's an exotic-animal preserve that houses a group of four-year-old liger brothers named Hercules, Zeus, Vulcan, and Sinbad. Ligers, the offspring of a lion father and tiger mother, are the world's largest cats, weighing up to half a ton each — double the heft of either parent. They're hybrids, and you won't see them in accredited American zoos, which look askance at letting different species breed. But that's how it is with hybrids: they don't get much respect and they're easy to miss, even when they're right under your nose. .

And yet, when you start looking around, they're everywhere.

Zorses, wholphins, tigons, and beefaloes. Lepjags, zonkeys, camas, and bonanzees. These are some of the captive-bred mammalian hybrids that exist, and they're joined by a host of hybrid birds, fish, insects, and plants. Thanks to new techniques that allow scientists to isolate and compare DNA, more hybrids are turning up every year, and we're learning that some of them — such as the pizzly, a cross between a polar bear and a grizzly — can occur naturally in the wild.

Hybrids evoke wonder and fear, magic and folklore. Their very existence unsettles our concept of what's out there, now and in the past. In fact, scientists are currently debating the extent to which hybrid breeding may have occurred during the evolution of man. Some contend that interspecies hanky-panky between humans and chimps — resulting in, yes, "humanzees" — went on for a million years or more after the two species split off from a common ancestor. Even now, there may be ghostly traces of this forbidden genetic lambada in our chromosomes.

Sound hard to believe? That's the hybrid calling card. They strain credulity — even when they're staring you in the face.

At The Institute of Greatly Endangered and Rare Species (TIGERS), outside Myrtle Beach, the ligers share a fifty-acre spread with some eighty other nonhybrid cats, bears, primates, wolves, and raptors, as well as a white crocodile and an African elephant. The animal trainer Bhagavan Antle, forty-seven, runs TIGERS with a crew of assistants who live on the grounds, eat vegetarian meals, and learn how to work safely with the menagerie.

I first meet Antle at the fenced-off preserve on a cold, wet January day. Inside a safari-themed lodge used to greet visitors who pay for private tours, he shows me recordings of his numerous media appearances and movie gigs. He's provided animals for such Hollywood films as *Ace Ventura: When Nature Calls, Forrest Gump,* and *Dr. Dolittle.*

Antle, who wears a ponytail and hoop earrings, is something of a hybrid himself. Raised on an Arizona cattle ranch, in the late seventies he became a disciple of Swami Satchidananda, whose claim to fame was saying the opening blessing at Woodstock.

Antle's exotic-animal career just sort of happened after he started working at a health clinic affiliated with the swami's ashram, Yogaville, in Buckingham County, Virginia. In 1982, a visitor to the clinic gave Antle a tiger cub. Later, another visitor — one who worked for "tiger in your tank" Exxon — asked him to lecture on health issues, cub in tow, at a company gathering. By the mid-eighties, Antle had become a full-time exotic-animal guy, breeding and training large, charismatic species for exhibition and rental to the entertainment industry. To his astonishment, a few years later, his male lion Arthur successfully mated with one of his tigresses. A second liger litter arrived in 2002.

A movie that Antle had nothing to do with made ligers famous. In 2004's geek-glorifying *Napoleon Dynamite,* Napoleon sketches a liger in his school notebook and declares, "It's pretty much my favorite animal . . . Bred for its skills in magic." Antle, one of the few liger owners in the world, did the rounds with Anderson Cooper and Matt Lauer. "We had such a big splash of exposure," he says. "We saw the valuable public appeal. It was like opening a chapter of myth that had come to life."

After Antle finishes showing me around, three of his assistants appear on the lodge's deck with Sinbad. The supersize beast has lighter stripes than a tiger and a lion-shaped head with no mane. His arms look stubby and his pectoral muscles are sagging. As we watch through a glass wall, a woman offers a chunk of meat from atop a platform, to make Sinbad stand and show off his twelve-foot frame. The assistants guide him around using chains and a baby bottle, and then Antle invites me out for a closer look. He walks up and snuggles Sinbad's muzzle. "Hi, bud," he coos, as if he's playing kissy-face with a kitten.

Sinbad could remove Antle's head with a single chomp, but I'm more enchanted than scared. It's like seeing a Sasquatch or centaur in the flesh. "In our core belief, people don't want to accept the idea that two distinctly different-looking wild animals can reproduce," Antle says. "Ligers make people understand that hybridization is real."

Charles Darwin understood that hybridization is real, and it deeply confused him. In *The Origin of Species,* he devoted a chapter to hybrids, but their existence was a riddle he never really solved. Hybrid animals like mules, Darwin noted, are usually sterile. He deemed it a "strange arrangement" that nature would afford two species the "special power" to create hybrids but then prevent these offspring from propagating. He offered squishy theories about why this was so — nobody knew anything about genes then — and on how hybrids fit into his overarching theory of natural selection.

Nine years later, in a book that examined variation in domesticated animals, Darwin explored hybrids more closely — there was even a mention of ligers, which had first been bred in England in 1824. Darwin asserted that hybrids might inadvertently push back the evolutionary clock, resurrecting traits that were better left behind. He used the mixing of human racial groups as an example, stating that foreign travelers frequently remarked on "the degraded state and savage disposition of crossed races of man."

Darwin's hybrids-are-bad dictum became orthodoxy during evolutionary biology's "modern synthesis," in the 1930s and '40s, which firmly connected genetics to natural selection. The Harvard ornithologist Ernst Mayr, a leading neo-Darwinist, set the tone by dismissing hybrids as an evolutionary dead end.

Mayr's verdict involved a surprisingly contentious question: what exactly is a species? Darwin had seen it as an arbitrary designation for animals that have similar physical features. Mayr came up with a concrete definition known as the "biological species concept." A species, he declared, is a reproductively isolated group that can interbreed.

By this formula, a species was a fixed unit that was improved over time by forces like random mutation and mate selection, not by having "gene flow" — biologese for doin' the nasty — with other species. "Species were rocks," says Michael Arnold, an evolutionary biologist at the University of Georgia who studies hybridization in both plants and animals.

But maybe they aren't. Arnold is part of a growing camp that sees species as more liquid than solid, and he rejects the idea that hybrids are always evolutionary losers just because they often can't reproduce. "A lot of us have been hammering away on this for many years," says Arnold. "They used to call us the Mongol hordes at their gates, but now we're inside."

Arnold is pushing a profound reconceptualization of evolution, one in which hybrids are more than bit players. Forget the tree of life, with new species neatly branching off from a common ancestor. It's a web of life, and hybrids help genes flow in unexpected directions.

But what about their famous sterility? Some hybrids can reproduce, and Arnold stresses that rare events have an "overwhelming importance" in the evolutionary process. Hybrids often have desirable traits — some are more fit than either parent — and there have been instances when hybrids were able to find enough fertile hybrid partners to create a new species. This process appears to be under way right now in the United States, involving a hybrid of the Pecos pupfish and the sheepshead minnow that has greatly multiplied and expanded its range. And some scientists contend that matings between gray wolves and coyotes thousands of years ago created an entirely new species: the red wolf.

More commonly, though, hybrids mate with one of their parent species, influencing the mix of what gets passed along to subsequent generations; essentially, they provide a bridge for genes to cross the species divide.

In a paper about hybridization and primate evolution that Arnold cowrote last year for the journal *Zoology,* he offers several examples, including chimpanzees and bonobos. DNA studies suggest that these two great apes swapped genes sometime after separating from a shared ancestor at least 800,000 years ago. Arnold believes these ape cousins occasionally mated but that the resulting "bonanzees" did not establish a new species. Instead, they hooked up with either chimps or bonobos. Bonanzees ultimately vanished, but they left genetic footprints in the genomes of their descendants.

In addition to comparing genomes for evidence of unusual gene flow, scientists increasingly are using DNA analysis to confirm the existence of heretofore unknown natural hybrids, whose existence argues that this process still occurs. On April 16, 2006, a hunter in Canada's Northwest Territories shot a polar bear whose fur had an orangish tint. Research showed that this animal had a grizzly-bear father, making it the first confirmed wild pizzly ever found. (Pizzlies had been bred before in captivity.) In 2003 DNA analysis done by the U.S. Forest Service confirmed that five odd-looking felines found in Maine and Minnesota were bobcat-lynx hybrids, dubbed blynxes. Other DNA-confirmed hybrid mammals reported since 1999 include the forest/savanna elephant in sub-Saharan Africa, the mink-polecat in France, and a sheep-goat in Botswana.

How much, then, do hybrids contribute to evolution? Nobody really knows. That's what's so exciting about these new DNA discoveries: the story is still unfolding.

Leaving aside their evolutionary import, there's a simpler reason hybrids fascinate. Some are astonishingly beautiful.

In Ramona, California, Nancy Nunke raises zorses and zonkeys at a six-acre spread called the Spots 'N Stripes Ranch, which mainly exists to breed zebras and miniature horses for show and for sale to private animal owners. After I pass through a security gate, Nunke greets me at her house and then points out her one zorse and two zonkeys, who are peeping at us from nearby corrals.

Nunke introduces me to her zorse, Zantazia, which, at seven

months old, is still a zoal. This delicate creature has a sorrel coat, a horse's long and thin face, and white stripes on her head, neck, torso, and legs. The offspring of a quarter-horse mother and a Grevy's zebra father, she may end up standing taller than both. I reach out to stroke Zantazia's neck, but she backs away. Nunke says zorses and zonkeys are friendly, but they have to set the pace. "It's like if you walked down Main Street and someone threw his arms around you," she says. "You'd say, 'Hey, buddy, back off.'"

Nunke has a soft spot for all "stripeys," which she thinks are more playful and affectionate than horses. "Horses will rub on you because they have an itch," she says. "A zebra will rub on you because he's your best friend." Zorses inherit the souls of zebras, she adds. "If they have one stripe, you train them exactly like you train a zebra. The z is totally in them."

We walk over to meet the zonkey brothers, Zane and Zebediah, who have donkey faces and ears, caramel coats, and a dizzying array of black lines. "They're the most striped zonkeys in the world," Nunke boasts. "What a good boy," she says, patting Zane's striped neck. Zane brays, and he looks so much like a zebra that his *hee-haw* startles me.

Nunke says some purists believe it's wrong to breed hybrids, that they pollute the natural order. "Here's how I feel about it," she tells me, a bit of swagger in her voice. "Whatever God didn't want to cross, he didn't make genetically capable of crossing."

Of all the hybrids that are theoretically possible, none shocks the mind more than a cross between humans and apes, and there has been at least one attempt to create a humanzee. In 1910 Ilya Ivanovich Ivanov, a Russian pioneer in artificial insemination, proposed seeding a female chimpanzee with human sperm. At a zoology conference in Austria, he noted (oh-so-quaintly) that this method would avoid the ethical dilemma of forcing the two species to actually have sex. Sixteen years later, with the backing of the Soviet Academy of Sciences, Ivanov traveled to Africa and gave it a try.

The details of Ivanov's experiments came to light only in 2002, when a Russian science historian, Kirill Rossiianov, produced a thirty-nine-page paper about the work. The study was published in English in the journal *Science in Context*, and when I came across it I was astonished. I struck up an e-mail correspondence with

Rossiianov, then, late last year, met with him in Moscow. We spoke at a teashop around the corner from Red Square, the outré topic making me feel as if we were Cold War spies trading state secrets.

Over a ten-year period, Rossiianov was able to unearth Ivanov's diaries and lab notes from the Soviet archives. According to Rossiianov, Ivanov and his son, a biochemistry student, set up a lab at the botanical gardens near Conakry, French Guinea. On the morning of February 28, 1927, they wrapped two female chimps in nets and inseminated them with sperm from a local man. On June 25 they inseminated another chimp with human sperm, this time using a special cage and knocking her out with ethyl chloride. None of the three became pregnant.

Rossiianov, a shy man, told me Ivanov's work repulsed him. "What do you think about the ethical dimension of Ivanov's experiments?" he asked. "Because, I dare say, I found them disgustful. Even now I find it terrible difficult to understand."

And Ivanov had plans to take things further. He also asked Soviet authorities for permission to impregnate women in his own country with sperm from an orangutan named Tarzan, who lived at a primate station in the republic of Georgia. He got a green light, and at least one volunteer came forward, but Tarzan died before any tests took place. Ivanov, convicted of counterrevolutionary activities unrelated to these experiments, was sent to the gulag in 1930, ending his career.

So we still don't know whether humans and chimps could successfully hybridize, but it may have happened in the distant past. A paper published in *Nature* last year offers compelling evidence that our ancestors had prolonged sexual relations with chimpanzees. The study, led by the geneticist David Reich, of Harvard Medical School, compares large stretches of DNA from humans and chimps.

The researchers, who contend that the two species diverged from a common forebear about 5.4 million years ago, found that the chimpanzee and human X chromosomes are more similar to each other than they are to any other chromosomes. The best explanation, they suggest, is that matings between chimps and early humans would have produced fertile female hybrids, who then mated with chimps themselves and had similar-enough X chromosomes to produce fertile male hybrids. They estimate that this went on for

1.2 million years after the initial split between the species. Eventually, only humans mated with the hybrids, and the hybrids disappeared, leaving behind nothing but genetic traces in our chromosomes.

Bhagavan Antle, the liger trainer, keeps pretty busy — his many gigs include working parties and performing at a venerable Miami theme park called Parrot Jungle Island, where he displays a liger and a gigantic "crocosaurus," a saltwater/Siamese crocodile. This entails a lot of animal shuttling, and Antle invited me to join him on a road trip from Myrtle Beach to Miami, where he would take a liger to a fundraiser at an exotic cat sanctuary and, later, a Super Bowl bash.

We meet at the Myrtle Beach facility, where Antle and his team lead Hercules the liger and two tigers into a trailer with small windows. Antle and I ride in an RV; joining us are three assistants and a diapered nine-month-old orangutan named Apsara, who's a hybrid, too. (She's a blend of Bornean and Sumatran orangutans, which are different species.) The infant, which has a comical mess of wild orange hair, rides in a baby sling worn by an assistant. Except for the occasional *meep meep*, you wouldn't know we're rolling with an orang.

During the two-day trip to Miami, a truck-and-trailer hauling the big cats is always right behind us. We refuel at crowded truck stops, pull in to strip malls to buy groceries, and even park one night behind a Holiday Inn. No one notices any of the exotic animals until we're stopped at the agricultural checkpoint at the Florida state line. The officer, a good ol' boy with slick hair and big sideburns, checks the paperwork and stumbles on the word *liger.* "What's that?" he asks.

"A mix of lion and tiger," Antle says.

"Those exist?"

Antle nods and takes out his business card, which shows him sitting with three tigers and Jay Leno. "Jay *Leno!*" says the officer. "That beats all!"

He waves us on without bothering to peek inside the trailer at one of the rarest creatures in the world. I guess Jay Leno is pretty much his favorite animal.

JOHN COLAPINTO

The Interpreter

FROM *The New Yorker*

ONE MORNING LAST JULY, in the rain forest of northwestern Brazil, Dan Everett, an American linguistics professor, and I stepped from the pontoon of a Cessna floatplane onto the beach bordering the Maici River, a narrow, sharply meandering tributary of the Amazon. On the bank above us were some thirty people — short, dark-skinned men, women, and children — some clutching bows and arrows, others with infants on their hips. The people, members of a hunter-gatherer tribe called the Pirahã, responded to the sight of Everett — a solidly built man of fifty-five with a red beard and the booming voice of a former evangelical minister — with a greeting that sounded like a profusion of exotic songbirds, a melodic chattering scarcely discernible to the uninitiated as human speech. Unrelated to any other extant tongue, and based on just eight consonants and three vowels, Pirahã has one of the simplest sound systems known. Yet it possesses such a complex array of tones, stresses, and syllable lengths that its speakers can dispense with their vowels and consonants altogether and sing, hum, or whistle conversations. It is a language so confounding to nonnatives that until Everett and his wife, Keren, arrived among the Pirahã as Christian missionaries in the 1970s, no outsider had succeeded in mastering it. Everett eventually abandoned Christianity, but he and Keren have spent the past thirty years, on and off, living with the tribe, and in that time they have learned Pirahã as no other Westerners have.

"*Xaói hi gáisai xigíaihiabisaoaxái ti xabiíhai hiatíihi xigío hoíhi,*" Everett said in the tongue's choppy staccato, introducing me as someone who would be "staying for a short time" in the village. The men and women answered in an echoing chorus, *"Xaói hi goó kaisigíaihí xapagáiso."*

Everett turned to me. "They want to know what you're called in 'crooked head.'"

"Crooked head" is the tribe's term for any language that is not Pirahã, and it is a clear pejorative. The Pirahã consider all forms of human discourse other than their own to be laughably inferior, and they are unique among Amazonian peoples in remaining monolingual. They playfully tossed my name back and forth among themselves, altering it slightly with each reiteration, until it became an unrecognizable syllable. They never uttered it again but instead gave me a lilting Pirahã name: Kaaxáoi, that of a Pirahã man from a village downriver, whom they thought I resembled. "That's completely consistent with my main thesis about the tribe," Everett told me later. "They reject everything from outside their world. They just don't want it, and it's been that way since the day the Brazilians first found them in this jungle in the 1700s."

Everett, who this past fall became the chairman of the Department of Languages, Literature, and Cultures at Illinois State University, has been publishing academic books and papers on the Pirahã (pronounced pee-da-HAN) for more than twenty-five years. But his work remained relatively obscure until early in 2005, when he posted on his website an article titled "Cultural Constraints on Grammar and Cognition in Pirahã," which was published that fall in the journal *Cultural Anthropology*. The article described the extreme simplicity of the tribe's living conditions and culture. The Pirahã, Everett wrote, have no numbers, no fixed color terms, no perfect tense, no deep memory, no tradition of art or drawing, and no words for "all," "each," "every," "most," or "few" — terms of quantification believed by some linguists to be among the common building blocks of human cognition. Everett's most explosive claim, however, was that Pirahã displays no evidence of recursion, a linguistic operation that consists of inserting one phrase inside another of the same type, as when a speaker combines discrete thoughts ("the man is walking down the street," "the man is wearing a top hat") into a single sentence ("The man who is wearing a top hat is walking down the street"). Noam Chomsky, the influential linguistic theorist, has recently revised his theory of universal grammar, arguing that recursion is the cornerstone of all languages and is possible because of a uniquely human cognitive ability.

Steven Pinker, the Harvard cognitive scientist, calls Everett's paper "a bomb thrown into the party." For months, it was the subject of passionate debate on social-science blogs and Listservs. Everett, once a devotee of Chomskyan linguistics, insists not only that Pirahã is a "severe counterexample" to the theory of universal grammar but also that it is not an isolated case. "I think one of the reasons that we haven't found other groups like this," Everett said, "is because we've been told, basically, that it's not possible." Some scholars were taken aback by Everett's depiction of the Pirahã as a people of seemingly unparalleled linguistic and cultural primitivism. "I have to wonder whether he's some Borgesian fantasist, or some Margaret Mead being stitched up by the locals," one reader wrote in an e-mail to the editors of a popular linguistics blog.

I had my own doubts about Everett's portrayal of the Pirahã shortly after I arrived in the village. We were still unpacking when a Pirahã boy, who appeared to be about eleven years old, ran out from the trees beside the river. Grinning, he showed off a surprisingly accurate replica of the floatplane we had just landed in. Carved from balsa wood, the model was four feet long and had a tapering fuselage, wings, and pontoons, as well as propellers, which were affixed with small pieces of wire so that the boy could spin the blades with his finger. I asked Everett whether the model contradicted his claim that the Pirahã do not make art. Everett barely glanced up. "They make them every time a plane arrives," he said. "They don't keep them around when there aren't any planes. It's a chain reaction, and someone else will do it, but then eventually it will peter out." Sure enough, I later saw the model lying broken and dirty in the weeds beside the river. No one made another one during the six days I spent in the village.

In the wake of the controversy that greeted his paper, Everett encouraged scholars to come to the Amazon and observe the Pirahã for themselves. The first person to take him up on the offer was a forty-three-year-old American evolutionary biologist named Tecumseh Fitch, who in 2002 coauthored an important paper with Chomsky and Marc Hauser, an evolutionary psychologist and biologist at Harvard, on recursion. Fitch and his cousin Bill, a som-melier based in Paris, were due to arrive by floatplane in the Pirahã village a couple of hours after Everett and I did. As the plane landed on the water, the Pirahã, who had gathered at the river, be-gan to cheer. The two men stepped from the cockpit, Fitch toting a laptop computer into which he had programmed a week's worth of linguistic experiments that he intended to perform on the Pirahã. They were quickly surrounded by curious tribe members. The Fitch cousins, having traveled widely together to remote parts of the world, believed that they knew how to establish an instant rap-port with indigenous peoples. They brought their cupped hands to their mouths and blew loon calls back and forth. The Pirahã looked on stone-faced. Then Bill began to make a loud popping sound by snapping a finger of one hand against the opposite palm. The Pirahã remained impassive. The cousins shrugged sheepishly and abandoned their efforts.

"Usually you can hook people really easily by doing these funny little things," Fitch said later. "But the Pirahã kids weren't buying it, and neither were their parents." Everett snorted. "It's not part of their culture," he said. "So they're not interested."

A few weeks earlier, I had called Fitch in Scotland, where he is a professor at the University of St. Andrews. "I'm seeing this as an ex-ploratory fact-finding trip," he told me. "I want to see with my own eyes how much of this stuff that Dan is saying seems to check out."

Everett is known among linguistics experts for orneriness and an impatience with academic decorum. He was born into a working-class family in Holtville, a town on the California-Mexico border, where his hard-drinking father, Leonard, worked variously as a bar-tender, a cowboy, and a mechanic. "I don't think we had a book in the house," Everett said. "To my dad, people who taught at colleges and people who wore ties were 'sissies' — all of them. I suppose some of that is still in me." Everett's chief exposure to intellectual life was through his mother, a waitress, who died of a brain aneu-rysm when Everett was eleven. She brought home Reader's Digest condensed books and a set of medical encyclopedias, which Everett attempted to memorize. In high school, he saw the movie *My Fair Lady* and thought about becoming a linguist, because, he later wrote, Henry Higgins's work "attracted me intellectually, and be-cause it looked like phoneticians could get rich."

As a teenager, Everett played the guitar in rock bands (his keyboardist later became an early member of Iron Butterfly) and smoked pot and dropped acid, until the summer of 1968, when he met Keren Graham, another student at El Capitan High School in Lakeside. The daughter of Christian missionaries, Keren was brought up among the Satere people in northeastern Brazil. She invited Everett to church and brought him home to meet her family. "They were loving and caring and had all these groovy experiences in the Amazon," Everett said. "They supported me and told me how great I was. This was just not what I was used to." On October 4, 1968, at the age of seventeen, he became a born-again Christian. "I felt that my life had changed completely, that I had stepped from darkness into light — all the expressions you hear." He stopped using drugs, and when he and Keren were eighteen they married. A year later, the first of their three children was born, and they began preparing to become missionaries.

In 1976, after graduating with a degree in foreign missions from the Moody Bible Institute of Chicago, Everett enrolled with Keren in the Summer Institute of Linguistics, known as SIL, an international evangelical organization that seeks to spread God's word by translating the Bible into the languages of preliterate societies. They were sent to Chiapas, Mexico, where Keren stayed in a hut in the jungle with the couple's children — by this time, there were three — while Everett underwent grueling field training. He endured fifty-mile hikes and survived for several days deep in the jungle with only matches, water, a rope, a machete, and a flashlight.

The couple were given lessons in translation techniques, for which Everett proved to have a gift. His friend Peter Gordon, a linguist at Columbia University who has published a paper on the absence of numbers in Pirahã, says that Everett regularly impresses academic audiences with a demonstration in which he picks from among the crowd a speaker of a language that he has never heard. "Within about twenty minutes, he can tell you the basic structure of the language and how its grammar works," Gordon said. "He has incredible breadth of knowledge, is really, really smart, knows stuff inside out." Everett's talents were obvious to the faculty at SIL, who for twenty years had been trying to make progress in Pirahã, with little success. In October 1977, at SIL's invitation, Everett, Keren, and their three small children moved to Brazil, first to the city of Belém to learn Portuguese, and then, a year later, to a Pirahã village at the mouth of the Maici River. "At that time, we didn't know that Pirahã was linguistically so hard," Keren told me.

There are about 350 Pirahã spread out in small villages along the Maici and Marmelos rivers. The village that I visited with Everett was typical: seven huts made by propping palm-frond roofs on top of four sticks. The huts had dirt floors and no walls or furniture, except for a raised platform of thin branches to sleep on. These fragile dwellings, in which a family of three or four might live, lined a path that wound through low brush and grass near the riverbank. The people keep few possessions in their huts — pots and pans, a machete, a knife — and make no tools other than scraping implements (used for making arrowheads), loosely woven palm-leaf bags, and wood bows and arrows. Their only ornaments are simple necklaces made from seeds, teeth, feathers, beads, and soda-can pull-tabs, which they often get from traders who barter with the Pirahã for Brazil nuts, wood, and *sorva* (a rubbery sap used to make chewing gum), and which the tribe members wear to ward off evil spirits.

Unlike other hunter-gatherer tribes of the Amazon, the Pirahã have resisted efforts by missionaries and government agencies to teach them farming. They maintain tiny, weed-infested patches of ground a few steps into the forest, where they cultivate scraggly manioc plants. "The stuff that's growing in this village was either planted by somebody else or it's what grows when you spit the seed out," Everett said to me one morning as we walked through the village. Subsisting almost entirely on fish and game, which they catch and hunt daily, the Pirahã have ignored lessons in preserving meats by salting or smoking, and they produce only enough manioc flour to last a few days. (The Kawahiv, another Amazonian tribe that Everett has studied, make enough to last for months.) One of their few concessions to modernity is their dress: the adult men wear T-shirts and shorts that they get from traders; the women wear plain cotton dresses that they sew themselves.

"For the first several years I was here, I was disappointed that I hadn't gone to a 'colorful' group of people," Everett told me. "I thought of the people in the Xingu, who paint themselves and use the lip plates and have the festivals. But then I realized that this is the most intense culture that I could ever have hoped to experience. This is a culture that's invisible to the naked eye but that is incredibly powerful, the most powerful culture of the Amazon. Nobody has resisted change like this in the history of the Amazon, and maybe of the world."

According to the best guess of archaeologists, the Pirahã arrived in the Amazon between 10,000 and 40,000 years ago, after bands of *Homo sapiens* from Eurasia migrated to the Americas over the Bering Strait. The Pirahã were once part of a larger Indian group called the Mura but had split from the main tribe by the time the Brazilians first encountered the Mura, in 1714. The Mura went on to learn Portuguese and to adopt Brazilian ways, and their language is believed to be extinct. The Pirahã, however, retreated deep into the jungle. In 1921 the anthropologist Curt Nimuendajú spent time among the Pirahã and noted that they showed "little interest in the advantages of civilization" and displayed "almost no signs of permanent contact with civilized people."

SIL first made contact with the Pirahã nearly fifty years ago, when a missionary couple, Arlo and Vi Heinrichs, joined a settlement on the Marmelos. The Heinrichses stayed for six and a half years, struggling to become proficient in the language. The phonemes (the sounds from which words are constructed) were exceedingly difficult, featuring nasal whines and sharp intakes of breath and sounds made by popping or flapping the lips. Individual words were hard to learn, since the Pirahã habitually whittle nouns down to single syllables. Also confounding was the tonal nature of the language: the meanings of words depend on changes in pitch. (The words for "friend" and "enemy" differ only in the pitch of a single syllable.) The Heinrichses' task was further complicated because Pirahã, like a few other Amazonian tongues, has male and female versions: the women use one fewer consonant than the men do.

"We struggled even getting to the place where we felt comfortable with the beginning of a grammar," Arlo Heinrichs told me. It was two years before he attempted to translate a Bible story; he chose the Prodigal Son from the book of Luke. Heinrichs read his halting translation to a Pirahã male. "He kind of nodded and said, in his way, 'That's interesting,'" Heinrichs recalled. "But there was no spiritual understanding — it had no emotional impact. It was just a story." After suffering repeated bouts of malaria, the couple were reassigned by SIL to administrative jobs in the city of Brasília, and in 1967 they were replaced with Steve Sheldon and his wife, Linda.

Sheldon earned a master's degree in linguistics during the time he spent with the tribe, and he was frustrated that Pirahã refused to conform to expected patterns. As he and his wife complained in workshops with SIL consultants. "We would say, 'It just doesn't seem that there's any way that it does X, Y, or Z,'" Sheldon recalled. "And the standard answer — since this typically doesn't happen in languages — was 'Well, it must be there, just look a little harder.'" Sheldon's anxiety over his slow progress was acute. He began many mornings by getting sick to his stomach. In 1977, after spending ten years with the Pirahã, he was promoted to director of SIL in Brazil and asked the Everetts to take his place in the jungle.

Everett and his wife were welcomed by the villagers, but it was months before they could conduct a simple conversation in Pirahã. "There are very few places in the world where you have to learn a language with no language in common," Everett told me. "It's called a monolingual field situation." He had been trained in the technique by his teacher at SIL, the late Kenneth L. Pike, a legendary field linguist and the chairman of the linguistics department at the University of Michigan. Pike, who created a method of language analysis called tagmemics, taught Everett to start with common nouns. "You find out the word for 'stick,'" Everett said. "Then you try to get the expression for 'two sticks' and for 'one stick drops to the ground,' 'two sticks drop to the ground.' You have to act everything out, to get some basic notion of how the clause structure works — where the subject, verb, and object go."

The process is difficult, as I learned early in my visit with the Pirahã. One morning, while applying bug repellent, I was watched by an older Pirahã man, who asked Everett what I was doing. Eager to communicate with him in sign language, I pressed together the thumb and index finger of my right hand and weaved them through the air while making a buzzing sound with my mouth. Then I brought my fingers to my forearm and slapped the spot where my fingers had alighted. The man looked puzzled and said to Everett, "He hit himself." I tried again — this time making a more insistent buzzing. The man said to Everett, "A plane landed on his arm." When Everett explained to him what I was doing, the man studied me with a look of pitying contempt, then turned away. Everett laughed. "You were trying to tell him something about your general state — that bugs bother you," he said. "They never talk that way, and they could never understand it. Bugs are a part of life."

"OK," I said. "But I'm surprised he didn't know I was imitating an insect."

"Think of how cultural that is," Everett said. "The movement of your hand. The sound. Even the way we represent animals is cultural."

Everett had to bridge many such cultural gaps in order to gain more than a superficial grasp of the language. "I went into the jungle, helped them make fields, went fishing with them," he said. "You cannot become one of them, but you've got to do as much as you can to feel and absorb the language." The tribe, he maintains, has no collective memory that extends back more than one or two generations and no original creation myths. Marco Antonio Gonçalves, an anthropologist at the Federal University of Rio de Janeiro, spent eighteen months with the Pirahã in the 1980s and wrote a dissertation on the tribe's beliefs. Gonçalves, who spoke limited Pirahã, agrees that the tribe has no creation myths but argues that few Amazonian tribes do. When pressed about what existed before the Pirahã and the forest, Everett says, the tribespeople invariably answer, "It has always been this way."

Everett also learned that the Pirahã have no fixed words for colors and instead use descriptive phrases that change from one moment to the next. "So if you show them a red cup, they're likely to say, 'This looks like blood,'" Everett said. "Or they could say, 'This is like *vrvcum*' — a local berry that they use to extract a red dye."

By the end of their first year, Dan Everett had a working knowledge of Pirahã. Keren tutored herself by strapping a cassette recorder around her waist and listening to audiotapes while she performed domestic tasks. (The Everetts lived in a thatched hut that was slightly larger and more sophisticated than the huts of the Pirahã; it had walls and a storage room that could be locked.)

During the family's second year in the Amazon, Keren and the Everetts' eldest child, Shannon, contracted malaria, and Keren lapsed into a coma. Everett borrowed a boat from river traders and trekked through the jungle for days to get her to a hospital. As soon as she was discharged, Everett returned to the village. (Keren recuperated in Belém for several months before joining him.) "Christians who believe in the Bible believe that it is their job to bring others the joy of salvation," Everett said. "Even if they're murdered, beaten to death, imprisoned — that's what you do for God."

Until Everett arrived in the Amazon, his training in linguistics had been limited to field techniques. "I wanted as little formal linguistic theory as I could get by with," he told me. "I wanted the basic linguistic training to do a translation of the New Testament." This changed when SIL lost its contract with the Brazilian government to work in the Amazon. SIL urged the Everetts to enroll as graduate students at the State University of Campinas (UNICAMP), in the state of São Paulo, since the government would give them permission to continue living on tribal lands only if they could show that they were linguists intent on recording an endangered language. At UNICAMP, in the fall of 1978, Everett discovered Chomsky's theories. "For me, it was another conversion experience," he said.

In the late 1950s, when Chomsky, then a young professor at MIT, first began to attract notice, behaviorism dominated the social sciences. According to B. F. Skinner, children learn words and grammar by being praised for correct usage, much as lab animals learn to push a lever that supplies them with food. In 1959, in a demolishing review of Skinner's book *Verbal Behavior,* Chomsky wrote that the ability of children to create grammatical sentences that they have never heard before proves that learning to speak does not depend on imitation, instruction, or rewards. As he put it in his book *Reflections on Language* (1975), "To come to know a human language would be an extraordinary intellectual achievement for a creature not specifically designed to accomplish this task."

Chomsky hypothesized that a specific faculty for language is encoded in the human brain at birth. He described it as a "language organ," which is equipped with an immutable set of rules — a universal grammar — that is shared by all languages, regardless of how different they appear to be. The language organ, Chomsky said, cannot be dissected in the way that a liver or a heart can, but it can be described through detailed analyses of the abstract structures underlying language. "By studying the properties of natural languages, their structure, organization, and use," Chomsky wrote, "we may hope to gain some understanding of the specific characteristics of human intelligence. We may hope to learn something about human nature."

Beginning in the 1950s, Chomskyans at universities around the world engaged in formal analyses of language, breaking sentences down into ever more complex tree diagrams that showed branching noun, verb, and prepositional phrases and also "X-bars," "transformations," "movements," and "deep structures" — Chomsky's terms for some of the elements that constitute the organizing principles of all language. "I'd been doing linguistics at a fairly low level of rigor," Everett said. "As soon as you started reading Chomsky's stuff, and the people most closely associated with Chomsky, you realized this is a totally different level — this is actually something that looks like science." Everett conceived his Ph.D. dissertation at UNICAMP as a strict Chomskyan analysis of Pirahã. Dividing his time between São Paulo and the Pirahã village, where he collected data, Everett completed his thesis in 1983. Written in Portuguese and later published as a book in Brazil, *The Pirahã Language and the Theory of Syntax* was a highly technical discussion replete with Chomskyan tree diagrams. However, Everett says that he was aware that Pirahã contained many linguistic anomalies that he could not fit into Chomsky's paradigm. "I knew I was leaving out a lot of stuff," Everett told me. "But these gaps were unexplainable to me."

The dissertation earned Everett a fellowship from the American Council of Learned Societies and a grant from the National Science Foundation to spend the 1984–85 academic year as a visiting fellow at MIT. Everett occupied an office next to Chomsky's; he found the famed professor brilliant but withering. "Whenever you try out a theory on someone, there's always some question that you hope they won't ask," Everett said. "That was always the first thing Chomsky would ask."

In 1988 Everett was hired by the University of Pittsburgh. By then, Chomsky's system of rules had reached a state of complexity that even Chomsky found too baroque, and he had begun to formulate a simpler model for the principles underlying all languages. Everett faithfully kept abreast of these developments. "Chomsky sent me all the papers that he was working on," he said. "I was like many of the scholars in that I made regular pilgrimages to sit in Chomsky's classes to collect the handouts and to figure out exactly where the theory was today." At the same time, Everett says that he was increasingly troubled by the idiosyncrasies of Pirahã. "None of it was addressed by Chomskyan linguistics," he told me. "Chomsky's theory only allows you to talk about properties that obtain of tree structures."

In the early 1990s, Everett began to reread the work of linguists who had preceded Chomsky, including that of Edward Sapir, an influential Prussian-born scholar who died in 1939. A student of the anthropologist Franz Boas, Sapir had taught at Yale and studied the languages of dozens of tribes in the Americas. Sapir was fascinated by the role of culture in shaping languages, and although he anticipated Chomsky's preoccupation with linguistic universals, he was more interested in the variations that made each language unique. In his 1921 book, *Language,* Sapir stated that language is an acquired skill, which "varies as all creative effort varies — not as consciously, perhaps, but nonetheless as truly as do the religions, the beliefs, the customs, and the arts of different peoples." Chomsky, however, believed that culture played little role in the study of language and that going to far-flung places to record the arcane babel of near-extinct tongues was a pointless exercise. Chomsky's view had prevailed. Everett began to wonder if this was an entirely good thing.

"When I went back and read the stuff Sapir wrote in the twenties, I just realized, hey, this really is a tradition that we lost," Everett said. "People believe they've actually studied a language when they have given it a Chomskyan formalism. And you may have given us absolutely no insight whatsoever into that language as a separate language."

Everett began to question the first principle of Chomskyan linguistics: that infants could not learn language if the principles of grammar had not been preinstalled in the brain. Babies are bathed in language from the moment they acquire the capacity to hear in the womb, Everett reasoned, and parents and caregivers expend great energy teaching children how to say words and assemble them into sentences — a process that lasts years. Was it really true that language, as Chomsky asserted, simply "grows like any other body organ"? Everett did not deny the existence of a biological endowment for language — humans couldn't talk if they did not possess the requisite neurological architecture to do so. But, con-

vinced that culture plays a far greater role than Chomsky's theory
accounted for, he decided that he needed to "take a radical reex-
amination of my whole approach to the problem."

In 1998, after nine years as chairman of the linguistics department
at the University of Pittsburgh, Everett became embroiled in a dispute with the new dean of the arts and sciences faculty. Keren was
completing a master's in linguistics at the university and was being
paid to work as a teaching assistant in Everett's department. Everett
was accused of making improper payments to Keren totaling some
$2,000, and he was subjected to an audit. He was exonerated, but
the allegation of misconduct infuriated him. Keren urged him to
quit his job so that they could return to the jungle and resume
their work as missionaries among the Pirahã.

It had been more than a decade since Everett had done any concerted missionary work — a reflection of his waning religious faith.
"As I read more and I got into philosophy and met a lot of friends
who weren't Christians, it became difficult for me to sustain the belief structure in the supernatural," he said. But he was inclined to
return to the Amazon, partly because he hoped to rekindle his
faith and partly because he was disillusioned with the theory that
had been the foundation of his intellectual life for two decades. "I
couldn't buy Chomsky's world-view any longer," Everett told me,
"and I began to feel that academics was a hollow and insignificant
way to spend one's life."

In the fall of 1999, Everett quit his job, and on the banks of the
Maici River he and Keren built a two-room, eight-meter-by-eight-
meter, bug- and snake-proof house from fourteen tons of ironwood
that he had shipped in by boat. Everett equipped the house with a
gas stove, a generator-driven freezer, a water-filtration system, a TV,
and a DVD player. "After twenty years of living like a Pirahã, I'd
had it with roughing it," he said. He threw himself into missionary work, translating the book of Luke into Pirahã and reading it
to tribe members. His zeal soon dissipated, however. Convinced
that the Pirahã assigned no spiritual meaning to the Bible, Everett
finally admitted that he did not, either. He declared himself an
atheist and spent his time tending house and studying linguistics.
In 2000, on a trip to Porto Velho, a town about two hundred miles
from the village, he found a month-old e-mail from a colleague at

the University of Manchester, inviting him to spend a year as a research professor at the school. In 2002 Everett was hired to a full-time position, and he and Keren moved to England. Three years later, he and Keren separated; she returned to Brazil, where she divides her time between the Pirahã village and an apartment in Porto Velho. He moved back to the United States in the fall of 2006 to begin a new job at Illinois State. Today Everett says that his three years in the jungle were hardly time wasted. "This new beginning with the Pirahã really was quite liberating," he told me. "Free from Chomskyan constraints, I was able to imagine new relationships between grammar and culture."

It is a matter of some vexation to Everett that the first article on the Pirahã to attract significant attention was written not by him but by his friend (and former colleague at the University of Pittsburgh) Peter Gordon, now at Columbia, who in 2004 published a paper in *Science* on the Pirahã's understanding of numbers. Gordon had visited the tribe with Everett in the early nineties, after Everett told him about the Pirahã's limited "one-two-many" counting system. Other tribes, in Australia, the South Sea Islands, Africa, and the Amazon, have a one-two-many numerical system, but with an important difference: they are able to learn to count in another language. The Pirahã have never been able to do this, despite concerted efforts by the Everetts to teach them to count to ten in Portuguese.

During a two-month stay with the Pirahã in 1992, Gordon ran several experiments with tribe members. In one he sat across from a Pirahã subject and placed in front of himself an array of objects — nuts, AA batteries — and had the Pirahã match the array. The Pirahã could perform the task accurately when the array consisted of two or three items, but their performance with larger groupings was, Gordon later wrote, "remarkably poor." Gordon also showed subjects nuts, placed them in a can, and withdrew them one at a time. Each time he removed a nut, he asked the subject whether there were any left in the can. The Pirahã answered correctly only with quantities of three or fewer. Through these and other tests, Gordon concluded that Everett was right: the people could not perform tasks involving quantities greater than three. Gordon ruled out mass retardation. Though the Pirahã do not allow marriage outside their tribe, they have long kept their gene pool refreshed by permitting women to sleep with outsiders. "Besides," Gordon said, "if there was some kind of Appalachian inbreeding or retardation going on, you'd see it in hairlines, facial features, motor ability. It bleeds over. They don't show any of that."

Gordon surmised that the Pirahã provided support for a controversial hypothesis advanced early in the last century by Benjamin Lee Whorf, a student of Sapir's. Whorf argued that the words in our vocabulary determine how we think. Since the Pirahã do not have words for numbers above two, Gordon wrote, they have a limited ability to work with quantities greater than that. "It's language affecting thought," Gordon told me. His paper, "Numerical Cognition Without Words: Evidence from Amazonia," was enthusiastically taken up by a coterie of "neo-Whorfian" linguists around the world.

Everett did not share this enthusiasm; in the ten years since he had introduced Gordon to the tribe, he had determined that the Pirahã have no fixed numbers. The word that he had long taken to mean "one" (*hoi,* on a falling tone) is used by the Pirahã to refer, more generally, to "a small size or amount," and the word for "two" (*hoi,* on a rising tone) is often used to mean "a somewhat larger size or amount." Everett says that his earlier confusion arose over what's known as the translation fallacy: the conviction that a word in one language is identical to a word in another, simply because in some instances they overlap in meaning. Gordon had mentioned the elastic boundaries of the words for "one" and "two" in his paper, but in Everett's opinion he had failed to explore the significance of the phenomenon. (Gordon disagrees, and for a brief period the two did not speak.)

Shortly after Gordon's article appeared, Everett began outlining a paper correcting what he believed were Gordon's errors. Its scope grew as Everett concluded that the Pirahã's lack of numerals was part of a larger constellation of "gaps." Over the course of three weeks, Everett wrote what would become his *Cultural Anthropology* article, 25,000 words in which he advanced a novel explanation for the many mysteries that had bedeviled him. Inspired by Sapir's cultural approach to language, he hypothesized that the tribe embodies a living-in-the-present ethos so powerful that it has affected every aspect of the people's lives. Committed to an existence in which only observable experience is real, the Pirahã do not think or speak in abstractions — and thus do not use color terms, quantifiers, numbers, or myths. Everett pointed to the word *xibípío* as a clue to how the Pirahã perceive reality solely according to what exists within the boundaries of their direct experience — which Everett defined as anything that they can see and hear or that someone living has seen and heard. "When someone walks around a bend in the river, the Pirahã say that the person has not simply gone away but *xibípío* — 'gone out of experience,'" Everett said. "They use the same phrase when a candle flame flickers. The light 'goes in and out of experience.'"

To Everett, the Pirahã's unswerving dedication to empirical reality — he called it the "immediacy-of-experience principle" — explained their resistance to Christianity, since the Pirahã had always reacted to stories about Christ by asking, "Have you met this man?" Told that Christ died 2,000 years ago, the Pirahã would react much as they did to my using bug repellent. It explained their failure to build up food stocks, since this required planning for a future that did not yet exist; it explained the failure of the boys' model airplanes to foster a tradition of sculpture-making, since the models expressed only the momentary burst of excitement that accompanied the sight of an actual plane. It explained the Pirahã's lack of original stories about how they came into being, since this was a conundrum buried in a past outside the experience of parents and grandparents.

Everett was convinced that the Pirahã's immediacy-of-experience principle went further still, "extending its tentacles," as he put it, "deep into their core grammar," to that feature that Chomsky claimed was present in all languages: recursion. Chomsky and other experts use the term to describe how we construct even the simplest utterances. "The girl jumped on the bed" is composed of a noun phrase ("the girl"), a verb ("jumped"), and a prepositional phrase ("on the bed"). In theory, as Chomsky has stressed, one could continue to insert chunks of language inside other chunks ad infinitum, thereby creating a never-ending sentence ("The man who is wearing a top hat that is slightly crushed around the brim although still perfectly elegant is walking down the street that was recently resurfaced by a crew of construction workers who tended to take coffee breaks that were a little too long while eating a hot dog that was . . .""). Or one could create sentences of never-ending variety. The capacity to generate unlimited meaning by placing one thought inside another is the crux of Chomsky's theory — what he calls, quoting the early-nineteenth-century German linguist Wilhelm von Humboldt, "the infinite use of finite means."

According to Everett, however, the Pirahã do not use recursion to insert phrases one inside another. Instead, they state thoughts in discrete units. When I asked Everett if the Pirahã could say, in their language, "I saw the dog that was down by the river get bitten by a snake," he said, "No. They would have to say, 'I saw the dog. The dog was at the beach. A snake bit the dog.'" Everett explained that because the Pirahã accept as real only that which they observe, their speech consists only of direct assertions ("The dog was at the beach"), and he maintains that embedded clauses ("that was down by the river") are not assertions but supporting, quantifying, or qualifying information — in other words, abstractions.

In his article, Everett argued that recursion is primarily a cognitive, not a linguistic, trait. He cited an influential 1962 article, "The Architecture of Complexity," by Herbert Simon, a Nobel Prize–winning economist, cognitive psychologist, and computer scientist, who asserted that embedding entities within like entities (in a recursive tree structure of the type central to Chomskyan linguistics) is simply how people naturally organize information. "Microsoft Word is organized by tree structures," Everett said. "You open up one folder and that splits into two other things, and that splits into two others. That's a tree structure. Simon argues that this is essential to the way humans organize information and is found in all human intelligence systems. If Simon is correct, there doesn't need to be any specific linguistic principle for this because it's just general cognition." Or, as Everett sometimes likes to put it: "The ability to put thoughts inside other thoughts is just the way humans are, because we're smarter than other species." Everett says that the Pirahã have this cognitive trait but that it is absent from their syntax because of cultural constraints.

Some scholars believe that Everett's claim that the Pirahã do not use recursion is tantamount to calling them stupid. Stephen Levinson, the neo-Whorfian director of the Language and Cognition Group at the Max Planck Institute for Psycholinguistics in the Netherlands, excoriated Everett in print for "having made the

Pirahã sound like the mindless bearers of an almost subhumanly simple culture." Anna Wierzbicka, a linguist at the Australian National University, was also troubled by the paper and told me, "I think from the point of view of — I don't know — human solidarity, human rights, and so on, it's really very important to know that it's a question that many people don't dare to raise, whether we have the same cognitive abilities or not, we humans."

Everett dismissed such criticisms, since he expressly states in the article that the unusual aspects of the Pirahã are not a result of mental deficiency. A Pirahã child removed from the jungle at birth and brought up in any city in the world, he said, would have no trouble learning the local tongue. Moreover, Everett pointed out, the Pirahã are supremely gifted in all the ways necessary to ensure their continued survival in the jungle: they know the usefulness and location of all important plants in their area; they understand the behavior of local animals and how to catch and avoid them; and they can walk into the jungle naked, with no tools or weapons, and walk out three days later with baskets of fruit, nuts, and small game. "They can outsurvive anybody, any other Indian in this region," he said. "They're very intelligent people. It never would occur to me that saying they lack things that Levinson or Wierzbicka predict they should have is calling them mindless idiots."

For Everett, the most important reaction to the article was Chomsky's. In an e-mail to Everett last April, Chomsky rejected Everett's arguments that the Pirahã's lack of recursion is a strong counterexample to his theory of universal grammar, writing, "UG is the true theory of the genetic component that underlies acquisition and use of language." He added that there is "no coherent alternative to UG." Chomsky declined to be interviewed for this article, but he referred me to "Pirahã Exceptionality: A Reassessment," a paper that was coauthored by David Pesetsky, a colleague of Chomsky's at MIT; Andrew Nevins, a linguist at Harvard; and Cilene Rodrigues, a linguist at UNICAMP. In the paper, which was posted last month on the website LingBuzz, a repository of articles on Chomskyan generative grammar, the authors used data from Everett's 1983 Ph.D. dissertation, as well as from a paper he published on Pirahã in 1986, to refute his recent claims about the language's unusual features — including the assertion that the Pirahã do not use recursion. The authors conceded that even in these early works, Everett had noted the absence of certain recursive structures in Pirahã. (The tribe, Everett wrote in the early eighties, does not embed possessives inside one another, as English speakers do when they say, "Tom's uncle's car's windshield . . ."). Nevertheless, they argued, Everett's early data suggested that the Pirahã's speech did contain recursive operations.

The fact that Everett had collected the data twenty-five years earlier, when he was a devotee of Chomsky's theory, was irrelevant, Pesetsky told me in an e-mail. At any rate, Pesetsky wrote, he and his coauthors detected "no sign of a particularly Chomskyan perspective" in the descriptive portions of Everett's early writings, adding, "For the most part, those works are about facts, and the categorizing of facts."

Everett, who posted a response to Pesetsky and his coauthors on LingBuzz, says that Chomsky's theory necessarily colored his data gathering and analysis. "'Descriptive work' apart from theory does not exist," he told me. "We ask the questions that our theories tell us to ask." In his response on LingBuzz, Everett addressed his critics' arguments point by point and disputed the contention that his early work was more reliable than his current research as a guide to Pirahã. "I would find the opposite troubling — i.e., that a researcher never changed their mind or found errors in their earlier work," he wrote. He added, "There are alternatives to Universal Grammar, and the fact that NPR" — Nevins, Pesetsky, and Rodrigues — "insist on characterizing the issue as though there were no alternatives, although typical, is either ignorant or purposely misleading."

In a comment on Everett's paper published in *Cultural Anthropology*, Michael Tomasello, the director of the Department of Developmental and Comparative Psychology at the Max Planck Institute for Evolutionary Anthropology, in Leipzig, endorsed Everett's conclusions that culture can shape core grammar. Because the Pirahã "talk about different things [than we do], different things get grammaticalized," he wrote, adding that "universal grammar was a good try, and it really was not so implausible at the time it was proposed, but since then we have learned a lot about many different languages, and they simply do not fit one universal cookie cutter."

The Harvard cognitive scientist Steven Pinker, who wrote admiringly about some of Chomsky's ideas in his 1994 bestseller, *The Language Instinct*, told me, "There's a lot of strange stuff going on in the Chomskyan program. He's a guru, he makes pronouncements that his disciples accept on faith and that he doesn't feel compelled to defend in the conventional scientific manner. Some of them become accepted within his circle as God's truth without really being properly evaluated, and, surprisingly for someone who talks about universal grammar, he hasn't actually done the spadework of seeing how it works in some weird little language that they speak in New Guinea."

Pinker says that his own doubts about the "Chomskyan program" increased in 2002, when Marc Hauser, Chomsky, and Tecumseh Fitch published their paper on recursion in *Science*. The authors wrote that the distinctive feature of the human faculty of language, narrowly defined, is recursion. Dogs, starlings, whales, porpoises, and chimpanzees all use vocally generated sounds to communicate with other members of their species, but none do so recursively, and thus none can produce complex utterances of infinitely varied meaning. "Recursion had always been an important part of Chomsky's theory," Pinker said. "But in Chomsky Mark II, or Mark III, or Mark VII, he all of a sudden said that the only thing unique to language is recursion. It's not just that it's the universal that has to be there; it's the magic ingredient that makes language possible."

In early 2005, Pinker and Ray Jackendoff, a linguistics professor at Tufts University, published a critique of Hauser, Chomsky, and Fitch's paper in the journal *Cognition*. "In my paper with Ray, we argue that if you just magically inject recursion into a chimpanzee you're not going to get a human who can put words together into phrases, label concepts with words, name things that happened decades ago or that may or may not happen decades in the future," Pinker said. "There's more to language than recursion." Pinker and Jackendoff, in a reference to Everett's research, cited Pirahã as an example of a language that has "phonology, morphology, syntax, and sentences" but no recursion. Pinker, however, was quick to tell me that the absence of recursion in one of the more than 6,000 known languages is not enough to disprove Chomsky's ideas. "If you had something that was present in 5,999 of the languages, and someone found one language that didn't have it — well, I think there may be some anthropologists who would say, 'This shows that there's no universals, that anything can happen,'" he said. "But more likely you'd say, 'Well, what's going on with that weird language?'"

Contemporary linguists have generally avoided speculation about how humans acquired language in the first place. Chomsky himself has long demonstrated a lack of interest in language origins and expressed doubt about Darwinian explanations. "It is perfectly safe to attribute this development to 'natural selection,'" Chomsky has written, "so long as we realize that there is no substance to this assertion, that it amounts to nothing more than a belief that there is some naturalistic explanation for these phenomena." Moreover, Chomsky's theory of universal grammar, which was widely understood to portray language as a complex system that arose fully formed in the brain, discouraged inquiry into how language developed. "This totally slams the door on the question," Brent Berlin, a cognitive anthropologist at the University of Georgia, told me. "It acts as if, in some inexplicable way, almost mysteriously, language is hermetically sealed from the conditions of life of the people who use it to communicate. But this is not some kind of abstract, beautiful, mathematical, symbolic system that is not related to real life."

Berlin believes that Pirahã may provide a snapshot of language at an earlier stage of syntactic development. "That's what Dan's work suggests," Berlin said of Everett's paper. "The plausible scenarios that we can imagine are ones that would suggest that early language looks something like the kind of thing that Pirahã looks like now."

Tecumseh Fitch, a tall, patrician man with long, pointed sideburns and a boyishly enthusiastic manner, owes his unusual first name to his ancestor, the Civil War general William Tecumseh Sherman. Fitch attended Brown University and earned a Ph.D. there. As a biologist with a special interest in animal communication, Fitch discovered that red deer possess a descended larynx, an anatomical feature that scientists had previously believed was unique to human beings and central to the development of speech. (The descended larynx has since been found in koalas, lions, tigers, jaguars, and leopards.) Fitch, eager to understand how humans acquired language, turned to linguistics and was surprised to learn that

Chomsky had written little about the question. But in 1999 Fitch happened to read an interview that Chomsky had given to *Spare Change News,* a newspaper for the homeless in Cambridge. "I read it and all the stuff he said about evolution was almost more than he's ever said in any published thing — and here it is in *Spare Change!*" Fitch said. "And he just made a few points that made me realize what he'd been getting at in a more enigmatic fashion in some of his previous comments." Fitch invited Chomsky to speak to a class that he was coteaching at Harvard on the evolution of language. Afterward they talked for several hours. A few months later, Chomsky agreed to collaborate with Fitch and Hauser on a paper that would attempt to pinpoint the features of language that are unique to humans and that allowed *Homo sapiens* to develop language. The authors compared animal and human communication, eliminating the aspects of vocalization that are shared by both, and concluded that one operation alone distinguished human speech: recursion. In the course of working on the article, Fitch grew sympathetic to Chomsky's ideas and became an articulate defender of the theory of universal grammar.

When Fitch and Everett met in Porto Velho in July, two days before heading into the jungle, they seemed, by tacit agreement, to be avoiding talk of Chomsky. But on the eve of our departure, while we were sitting by the pool at the Hotel Vila Rica, Everett mentioned two professors who, he said, were "among the three most arrogant people I've met."

"Who's the third?" Fitch asked.

"Noam," Everett said.

"No!" Fitch cried. "Given his status in science, Chomsky is the least arrogant man, the humblest great man, I've ever met."

Everett was having none of it. "Noam Chomsky thinks of himself as Aristotle!" he declared. "He has dug a hole for linguistics that it will take decades for the discipline to climb out of!"

The men argued for the next two hours, though by the time they parted for the night civility had been restored, and the détente was still holding when they met in the Pirahã village the next day and agreed to begin experiments the following morning.

At sunrise, a group of some twenty Pirahã gathered outside Everett's house. They were to be paid for their work as experimental subjects — with tobacco, cloth, farina, and machetes. "And, believe me," Everett said, "that's the only reason they're here. They have no interest in what we're doing. They're hunter-gatherers, and they see us just like fruit trees to gather from."

Fitch went out with Everett into the thick heat, carrying his laptop. The two men, trailed by the Pirahã, followed a narrow path through the low underbrush to Everett's office, a small hut, raised off the ground on four-foot-high stilts, at the edge of the jungle. Fitch placed his computer on the desk and launched a program that he had spent several weeks writing in preparation for this trip.

Fitch's experiments were based on the so-called Chomsky hierarchy, a system for classifying types of grammar, ranked in ascending order of complexity. To test the Pirahã's ability to learn one of the simplest types of grammar, Fitch had written a program in which grammatically correct constructions were represented by a male voice uttering one nonsense syllable (*mi* or *doh* or *ga*, for instance), followed by a female voice uttering a different nonsense syllable (*lee* or *ta* or *gee*). Correct constructions would cause an animated monkey head at the bottom of the computer screen to float to a corner at the top of the screen after briefly disappearing; incorrect constructions (any time one male syllable was followed by another male syllable or more than one female syllable) would make the monkey head float to the opposite corner. Fitch set up a small digital movie camera behind the laptop to film the Pirahã's eye movements. In the few seconds' delay before the monkey head floated to either corner of the screen, Fitch hoped that he would be able to determine, from the direction of the subjects' unconscious glances, if they were learning the grammar. The experiment, using different stimuli, had been conducted with undergraduates and monkeys, all of whom passed the test. Fitch told me that he had little doubt that the Pirahã would pass. "My expectation coming in here is that they're going to act just like my Harvard undergrads," he said. "They're going to do exactly what every other human has done, and they're going to get this basic pattern. The Pirahã are humans — humans can do this."

Fitch called for the first subject.

Everett stepped outside the hut and spoke to a short, muscular man with a bowl-shaped haircut and heavily calloused bare feet. The man entered the hut and sat down at the computer, which promptly crashed. Fitch rebooted. It crashed again.

"It's the humidity," Everett said.

Fitch finally got the computer working, but then the video camera seized up.

"Goddamn Chomskyan," Everett said. "Can't even run an experiment."

Eventually, Fitch got all the equipment running smoothly and started the experiment. It quickly became obvious that the Pirahã man was simply watching the floating monkey head and wasn't responding to the audio cues.

"It didn't look like he was doing premonitory looking," Fitch said. "Maybe ask him to point to where he thinks the monkey is going to go."

"They don't point," Everett said. Nor, he added, do they have words for right and left. Instead, they give directions in absolute terms, telling others to head "upriver" or "downriver," or "to the forest" or "away from the forest." Everett told the man to say whether the monkey was going upriver or downriver. The man said something in reply.

"What did he say?" Fitch asked.

"He said, 'Monkeys go to the jungle.'"

Fitch grimaced in frustration. "Well, he's not guessing with his eyes," he said. "Is there another way he can indicate?"

Everett again told the man to say whether the monkey was going upriver or down. The man made a noise of assent. Fitch resumed the experiment, but the man simply waited until the monkey moved. He followed it with his eyes, laughed admiringly when it came to a stop, then announced whether it had gone upriver or down.

After several minutes of this, Fitch said, on a rising note of panic, "If they fail in the recursion one — it's not recursion; I've got to stop saying that. I mean embedding. Because, I mean, if he can't get *this* —"

"This is typical Pirahã," Everett said soothingly. "This is new stuff, and they don't do new stuff."

"But when they're hunting they must have those skills of visual anticipation," Fitch said.

"Yeah," Everett said dryly. "But this is not a real monkey." He pointed at the grinning animated head bobbing on the screen.

"Fuck!" Fitch said. "If I'd had a joystick for him to *hunt* the monkey!" He paced a little, then said, "The crazy thing is that this is already more realistic than the experiments Aslin did with babies."

"Look," Everett said, "the cognitive issue here is the cultural impediment to doing new things. He doesn't know there's a pattern to recognize."

Everett dismissed the man and asked another Pirahã to come into the hut. A young man appeared, wearing a green-and-yellow 2002 Brazilian World Cup shirt, and sat at the computer. Everett told him to say whether the monkey was going to go upriver or downriver.

Fitch ran the experiment. The man smiled and pointed with his chin whenever the monkey head came to rest.

"The other idea," Fitch said, "is if we got a bunch of the kids, and whoever points first gets a lollipop."

"That's got an element of competition that they won't go for," Everett said.

The computer crashed. Convinced that there was a glitch in the software, Fitch picked up the machine and carried it back to the main house to make repairs.

"This is typical of field work in the Amazon, which is why most people don't do it," Everett said. "But the problem here is not cognitive; it's cultural." He gestured toward the Pirahã man at the table. "Just because we're sitting in the same room doesn't mean we're sitting in the same century."

By the next morning, Fitch had debugged his software, but other difficulties persisted. One subject, a man in blue nylon running shorts, ignored instructions to listen to the syllables and asked questions about the monkey head: "Is that rubber?" "Does this monkey have a spouse?" "Is it a man?" Another man fell asleep midtrial (the villagers had been up all night riotously talking and laughing — a common occurrence for a people who do not live by the clock). Meanwhile, efforts to get subjects to focus were hampered by the other tribe members, who had collected outside the hut and held loud conversations that were audible through the screened windows.

Steve Sheldon, Everett's predecessor in the Pirahã village, had told me of the challenges he faced in the late sixties when he did research on behalf of Brent Berlin and Paul Kay (an anthropologist and linguist at the University of California at Berkeley), who were collecting data about colors from indigenous peoples. Sheldon had concluded that the Pirahã tribe has fixed color terms — a view duly enshrined in Berlin and Kay's book *Basic Color Terms: Their Universality and Evolution* (1969). Only later did Sheldon realize that his data were unreliable. Told to question tribe members in isolation, Sheldon had been unable to do so because the tribe refused to be split up; members had eavesdropped on Sheldon's interviews and collaborated on answers. "Their attitude was 'Who cares what the color is?'" Sheldon told me. "'But we'll give him something because that's what he wants.'" (Today Sheldon endorses Everett's claim that the tribe has no fixed color terms.)

Sheldon said that the Pirahã's obstructionist approach to researchers is a defensive gesture. "They have been made fun of by outsiders because they do things differently," Sheldon told me. "With researchers who don't speak their language, they make fun, giving really bad information, totally wrong information sometimes."

On the third day, Fitch had figured out that he was being hindered by some of the same problems that Sheldon had faced. That morning, he tacked up bedsheets over the window screens and demanded that the tribe remain at a distance from the hut. (Several yards away, Fitch's cousin, Bill, entertained the group by playing Charlie Parker tunes on his iPod.) Immediately, the testing went better. One Pirahã man seemed to make anticipatory eye movements, although it was difficult to tell, because his eyes were hard to make out under the puffy lids, a feature typical of the men's faces. Fitch tried the experiment on a young woman with large, dark irises, but it was not clear that her few correct glances were anything but coincidental. "Lot of random looks," Everett muttered. "It's not obvious that they're getting it either way," Fitch said.

On the fourth day, Fitch seemed to hit pay dirt. The subject was a girl of perhaps sixteen. Focused, alert, and calm, she seemed to grasp the grammar, her eyes moving to the correct corner of the screen in advance of the monkey's head. Fitch was delighted and perhaps relieved; before coming to the Amazon, he had told me that the failure of a Pirahã to perform this task would be tantamount to "discovering a Sasquatch."

Fitch decided to test the girl on a higher level of the Chomsky hierarchy, a "phrase-structure grammar." He had devised a program in which correct constructions consisted of any number of male syllables followed by an equal number of female syllables. Hauser, Chomsky, and Fitch, in their 2002 paper, had stated that a phrase-structure grammar, which makes greater demands on memory and pattern recognition, represents the minimum foundation necessary for human language.

Fitch performed several practice trials with the girl to teach her the grammar. Then he and Everett stepped back to watch. "If this is working," Fitch said, "we could try to get NSF money. This could be big — even for psychology."

At the mention of psychology — a discipline that often emphasizes the influence of environment on behavior and thus is at a remove from Chomsky's views — Everett laughed. "Now he's beginning to see it my way!" he said.

The girl gazed at the screen and listened as the HAL-like computer voices flatly intoned the meaningless syllables. Fitch peered at the camera's viewfinder screen, trying to discern whether the girl's eye movements indicated that she understood the grammar. It was impossible to say. Fitch would have to take the footage back to Scotland, where it would be vetted by an impartial postdoc volunteer, who would "score" the images on a time line carefully synchronized to the soundtrack of the spoken syllables, so that Fitch could say without a doubt whether the subject's eyes had anticipated the monkey head or merely followed it. (Recently Fitch said that the data "look promising," but he declined to elaborate, pending publication of his results.)

That evening, Everett invited the Pirahã to come to his home to watch a movie: Peter Jackson's remake of *King Kong*. (Everett had discovered that the tribe loves movies that feature animals.) After nightfall, to the grinding sound of the generator, a crowd of thirty or so Pirahã assembled on benches and on the wooden floor of Everett's "Indian room," a screened-off section of his house where he confines the Pirahã, owing to their tendency to spit on the floor. Everett had made popcorn, which he distributed in a large bowl. Then he started the movie, clicking ahead to the scene in which Naomi Watts, reprising Fay Wray's role, is offered as a sacrifice by the tribal people of an unspecified South Seas island. The Pirahã shouted with delight, fear, laughter, and surprise — and when

Kong himself arrived, smashing through the palm trees, pandemonium ensued. Small children, who had been sitting close to the screen, jumped up and scurried into their mothers' laps; the adults laughed and yelled at the screen.

If Fitch's experiments were inconclusive on the subject of whether Chomsky's universal grammar applied to the Pirahã, Jackson's movie left no question about the universality of Hollywood film grammar. As Kong battled raptors and Watts dodged giant insects, the Pirahã offered a running commentary, which Everett translated: "Now he's going to fall!" "He's tired!" "She's running!" "Look. A centipede!" Nor were the Pirahã in any doubt about what was being communicated in the long, lingering looks that passed between gorilla and girl. "She is his spouse," one Pirahã said. Yet in their reaction to the movie Everett also saw proof of his theory about the tribe. "They're not generalizing about the character of giant apes," he pointed out. "They're reacting to the immediate action on the screen with direct assertions about what they see."

In Fitch's final two days of experiments, he failed to find another subject as promising as the sixteen-year-old girl. But he was satisfied with what he had been able to accomplish in six days in the jungle. "I think Dan's is an interesting and valid additional approach to add to the arsenal," Fitch told me after we had flown back to Porto Velho and were sitting beside the pool at the Hotel Vila Rica. "I think you need to look at something as complex as language from lots of different angles, and I think the angle he's arguing is interesting and deserves more work, more research. But as far as the Pirahã disproving universal grammar? I don't think anything I could have seen out there would have convinced me that that was ever anything other than just the wrong way to frame the problem."

On my final night in Brazil, I met Keren Everett in the gloomy lobby of the hotel. At fifty-five, she is an ageless, elfin woman with large dark eyes and waist-length hair pulled back from her face. She is trained in formal linguistics, but her primary interest in the Pirahã remains missionary. In keeping with the tenets of SIL, she does not proselytize or actively attempt to convert them; it is enough, SIL believes, to translate the Bible into the tribal tongue. Keren insists that she does not know the language well yet. "I still haven't cracked it," she said, adding that she thought she was "beginning to feel it for the first time, after twenty-five years."

The key to learning the language is the tribe's singing, Keren said: the way that the group can drop consonants and vowels altogether and communicate purely by variations in pitch, stress, and rhythm — what linguists call "prosody." I was reminded of an evening in the village when I had heard someone singing a clutch of haunting notes on a rising, then falling scale. The voice repeated the pattern over and over, without variation, for more than half an hour. I crept up to the edge of one of the Pirahã huts and saw that it was a woman, winding raw cotton onto a spool and intoning this extraordinary series of notes that sounded like a muted horn. A toddler played at her feet. I asked Everett about this, and he said something vague about how tribe members "sing their dreams." But when I described the scene to Keren she grew animated and explained that this is how the Pirahã teach their children to speak. The toddler was absorbing the lesson in prosody through endless repetition — an example, one might argue, of Edward Sapir's cultural theory of language acquisition at work.

"This language uses prosody much more than any other language I know of," Keren told me. "It's not the kind of thing that you can write and capture and go back to; you have to watch, and you have to feel it. It's like someone singing a song. You want to watch and listen and try to sing along with them. So I started doing that, and I began noticing things that I never transcribed, and things I never picked up when I listened to a tape of them, and part of it was the performance. So at that point I said, 'Put the tape recorders and notebooks away, focus on the person, watch them.' They give a lot of things using prosody that you never would have found otherwise. This has never been documented in any language I know." Aspects of Pirahã that had long confounded Keren became clear, she said. "I realized, Oh! That's what the subject-verb looks like, that's what the pieces of the clause and the time phrase and the object and the other phrases feel like. That was the beginning of a breakthrough for me. I won't say that I've broken it until I can creatively use the verbal structure — and I can't do it yet."

Keren says that Everett's frustration at realizing that they would have to "start all over again" with the language ultimately led to his decision to leave the Amazon in 2002 and return to academia. "He was diligent and he was trying to use his perspective and his training, and I watched the last year that we were together in the village — he just was, like, 'This is it. I'm out of here.' That was the year I started singing, and he said, 'Damn it if I'm going to learn to sing this language!' And he was out. It's torment. It is tormenting when you have a good mind and you can't crack it. I said, 'I don't care, we're missing something. We've got to look at it from a different perspective.'" Keren shook her head. "Pirahã has just always been out there defying every linguist that's gone out there, because you can't start at the segment level and go on. You're not going to find out anything, because they really can communicate without the syllables."

Later that day, when Everett drove me to the airport in Porto Velho, I told him about my conversation with Keren. He sighed. "Keren has made tremendous progress, and I'm sure she knows more about musical speech than I do at this point," he said. "There's probably several areas of Pirahã where her factual knowledge exceeds mine. But it's not all the prosody. That's the thing." Keren's perspective on Pirahã derives from her missionary impulses, he said. "It would be impossible for her to believe that we know the language, because that would mean that the word of God doesn't work."

Everett pulled into the airport parking lot. It was clear that talking about Keren caused him considerable pain. He did not want our conversation to end on a quarrel with her. He reminded me that his disagreement is with Chomsky.

"A lot of people's view of Chomsky is of the person in the lead on the jungle path," Everett had told me in the Pirahã village. "And if anybody's likely to find the way home it's him. So they want to stay as close behind him as possible. Other people say, 'Fuck that, I'm going to get on the river and take my canoe.'"

CHRISTOPHER J. CONSELICE

The Universe's Invisible Hand

FROM *Scientific American*

WHAT TOOK US SO LONG? Only in 1998 did astronomers discover that we had been missing nearly three-quarters of the contents of the universe, the so-called dark energy — an unknown form of energy that surrounds each of us, tugging at us ever so slightly, holding the fate of the cosmos in its grip, but to which we are almost totally blind. Some researchers, to be sure, had anticipated that such energy existed, but even they will tell you that its detection ranks among the most revolutionary discoveries in twentieth-century cosmology. Not only does dark energy appear to make up the bulk of the universe, but its existence, if it stands the test of time, will probably require the development of new theories of physics.

Scientists are just starting the long process of figuring out what dark energy is and what its implications are. One realization has already sunk in: although dark energy betrayed its existence through its effect on the universe as a whole, it may also shape the evolution of the universe's inhabitants: stars, galaxies, galaxy clusters. Astronomers may have been staring at its handiwork for decades without realizing it.

Ironically, the very pervasiveness of dark energy is what made it so hard to recognize. Dark energy, unlike matter, does not clump in some places more than others; by its very nature, it is spread smoothly everywhere. Whatever the location — be it in your kitchen or in intergalactic space — it has the same density, about 10–26 kilograms per cubic meter, equivalent to a handful of hydrogen atoms. All the dark energy in our solar system amounts to the mass of a small asteroid, making it an utterly inconsequential player in the dance of the planets. Its effects stand out only when viewed over vast distances and spans of time.

Since the days of the American astronomer Edwin Hubble, observers have known that all but the nearest galaxies are moving away from us at a rapid rate. This rate is proportional to distance: the more distant a galaxy is, the faster its recession. Such a pattern implied that galaxies are not moving through space in the conventional sense but are being carried along as the fabric of space itself stretches (see "Misconceptions about the Big Bang," by Charles H. Lineweaver and Tamara M. Davis, *Scientific American,* March 2005). For decades, astronomers struggled to answer the obvious follow-up question: how does the expansion rate change over time? They reasoned that it should be slowing down, as the inward gravitational attraction exerted by galaxies on one another should have counteracted the outward expansion.

The first clear observational evidence for changes in the expansion rate involved distant supernovae, massive exploding stars that can be used as markers of cosmic expansion, just as watching driftwood lets you measure the speed of a river. These observations made clear that the expansion was slower in the past than it is today and is therefore accelerating. More specifically, it had been slowing down but at some point underwent a transition and began speeding up (see "Surveying Space-Time with Supernovae," by Craig J. Hogan, Robert P. Kirshner, and Nicholas B. Suntzeff, *Scientific American,* January 1999, and "From Slowdown to Speedup," by Adam G. Riess and Michael S. Turner, *Scientific American,* February 2004). This striking result has since been cross-checked by independent studies of the cosmic microwave background radiation by, for example, the Wilkinson Microwave Anisotropy Probe (WMAP).

One possible conclusion is that different laws of gravity apply on supergalactic scales than on lesser ones, so that galaxies' gravity does not, in fact, resist expansion. But the more generally accepted hypothesis is that the laws of gravity are universal and that some form of energy, previously unknown to science, opposes and overwhelms galaxies' mutual attraction, pushing them apart ever faster. Although dark energy is inconsequential within our galaxy (let alone your kitchen), it adds up to the most powerful force in the cosmos.

Cosmic Sculptor

As astronomers have explored this new phenomenon, they have found that, in addition to determining the overall expansion rate of the universe, dark energy has long-term consequences for smaller scales. As you zoom in from the entire observable universe, the first thing you notice is that matter on cosmic scales is distributed in a cobweb-like pattern — a filigree of filaments, several tens of millions of light-years long, interspersed with voids of similar size. Simulations show that both matter and dark energy are needed to explain the pattern.

That finding is not terribly surprising, though. The filaments and voids are not coherent bodies like, say, planets. They have not detached from the overall cosmic expansion and established their own internal equilibrium of forces. Rather they are features shaped by the competition between cosmic expansion (and any phenomenon affecting it) and their own gravity. In our universe, neither player in this tug of war is overwhelmingly dominant. If dark energy were stronger, expansion would have won and matter would be spread out rather than concentrated in filaments. If dark energy were weaker, matter would be even more concentrated than it is.

The situation gets more complicated as you continue to zoom in and reach the scale of galaxies and galaxy clusters. Galaxies, including our own Milky Way, do not expand with time. Their size is controlled by an equilibrium between gravity and the angular momentum of the stars, gas, and other material that make them up; they grow only by accreting new material from intergalactic space or by merging with other galaxies. Cosmic expansion has an insignificant effect on them. Thus it is not at all obvious that dark energy should have had any say whatsoever in how galaxies formed. The same is true of galaxy clusters, the largest coherent bodies in the universe — assemblages of thousands of galaxies embedded in a vast cloud of hot gas and bound together by gravity.

Yet it now appears that dark energy may be the key link among several aspects of galaxy and cluster formation that not long ago appeared unrelated. The reason is that the formation and evolution of these systems is partially driven by interactions and mergers between galaxies, which in turn may have been driven strongly by dark energy.

To understand the influence of dark energy on the formation of galaxies, first consider how astronomers think galaxies form. Current theories are based on the idea that matter comes in two basic kinds. First, there is ordinary matter, whose particles readily interact with one another and, if electrically charged, with electromagnetic radiation. Astronomers call this type of matter "baryonic" in reference to its main constituent, baryons, such as protons and neutrons. Second, there is dark matter (which is distinct from dark energy), which makes up 85 percent of all matter and whose salient property is that it comprises particles that do not react with radiation. Gravitationally, dark matter behaves just like ordinary matter.

According to models, dark matter began to clump immediately after the big bang, forming spherical blobs that astronomers refer to as "halos." The baryons, in contrast, were initially kept from clumping by their interactions with one another and with radiation. They remained in a hot, gaseous phase. As the universe expanded, this gas cooled and the baryons were able to pack themselves together. The first stars and galaxies coalesced out of this cooled gas a few hundred million years after the big bang. They did not materialize in random locations but in the centers of the dark-matter halos that had already taken shape.

Since the 1980s a number of theorists have done detailed computer simulations of this process, including groups led by Simon D. M. White of the Max Planck Institute for Astrophysics in Garching, Germany, and Carlos S. Frenk of Durham University in England. They have shown that most of the first structures were small, low-mass dark-matter halos. Because the early universe was so dense, these low-mass halos (and the galaxies they contained) merged with one another to form larger-mass systems. In this way, galaxy construction was a bottom-up process, like building a dollhouse out of Lego bricks. (The alternative would have been a top-down process, in which you start with the dollhouse and smash it to make bricks.) My colleagues and I have sought to test these models by looking at distant galaxies and how they have merged over cosmic time.

Galaxy Formation Peters Out

Detailed studies indicate that a galaxy gets bent out of shape when it merges with another galaxy. The earliest galaxies we can see existed when the universe was about a billion years old, and many of these indeed appear to be merging. As time went on, though, the fusion of massive galaxies became less common. Between 2 billion and 6 billion years after the big bang — that is, over the first half of cosmic history — the fraction of massive galaxies undergoing a merger dropped from half to nearly nothing at all. Since then, the distribution of galaxy shapes has been frozen, an indication that smashups and mergers have become relatively uncommon.

In fact, fully 98 percent of massive galaxies in today's universe are either elliptical or spiral, with shapes that would be disrupted by a merger. These galaxies are stable and comprise mostly old stars, which tells us that they must have formed early and have remained in a regular morphological form for quite some time. A few galaxies are merging in the present day, but they are typically of low mass.

The virtual cessation of mergers is not the only way the universe has run out of steam since it was half its current age. Star formation, too, has been waning. Most of the stars that exist today were born in the first half of cosmic history, as first convincingly shown by several teams in the 1990s, including ones led by Simon J. Lilly, then at the University of Toronto; Piero Madau, then at the Space Telescope Science Institute; and Charles C. Steidel of the California Institute of Technology. More recently, researchers have learned how this trend occurred. It turns out that star formation in massive galaxies shut down early. Ever since the universe was half its current age, only lightweight systems have continued to create stars at a significant rate. This shift in the venue of star formation is called galaxy downsizing (see "The Midlife Crisis of the Cosmos," by Amy J. Barger, *Scientific American,* January 2005). It seems paradoxical. Galaxy formation theory predicts that small galaxies take shape first, and, as they amalgamate, massive ones arise. Yet the history of star formation shows the reverse: massive galaxies are initially the main stellar birthing grounds, then smaller ones take over.

Another oddity is that the buildup of supermassive black holes, found at the centers of galaxies, seems to have slowed down considerably. Such holes power quasars and other types of active galaxies, which are rare in the modern universe; the black holes in our galaxy and others are quiescent. Are any of these trends in galaxy evo-

lution related? Is it really possible that dark energy is the root
cause?

The Steady Grip of Dark Energy

Some astronomers have proposed that internal processes in galaxies, such as energy released by black holes and supernovae, turned
off galaxy and star formation. But dark energy has emerged as possibly a more fundamental culprit, the one that can link everything
together. The central piece of evidence is the rough coincidence in
timing between the end of most galaxy and cluster formation and
the onset of the domination of dark energy. Both happened when
the universe was about half its present age.

The idea is that up to that point in cosmic history, matter
was so dense that gravitational forces among galaxies dominated
over the effects of dark energy. Galaxies rubbed shoulders, inter-
acted with one another, and frequently merged. New stars formed
as gas clouds within galaxies collided, and black holes grew when
gas was driven toward the centers of these systems. As time pro-
gressed and space expanded, matter thinned out and its gravity
weakened, whereas the strength of dark energy remained constant
(or nearly so). The inexorable shift in the balance between the
two eventually caused the expansion rate to switch from decelera-
tion to acceleration. The structures in which galaxies reside were
then pulled apart, with a gradual decrease in the galaxy merger
rate as a result. Likewise, intergalactic gas was less able to fall
into galaxies. Deprived of fuel, black holes became more quiescent.

This sequence could perhaps account for the downsizing of the
galaxy population. The most massive dark-matter halos, as well as
their embedded galaxies, are also the most clustered; they reside in
close proximity to other massive halos. Thus they are likely to
knock into their neighbors earlier than are lower-mass systems.
When they do, they experience a burst of star formation. The
newly formed stars light up and then blow up, heating the gas and
preventing it from collapsing into new stars. In this way, star formation chokes itself off: stars heat the gas from which they emerged,
preventing new ones from forming. The black hole at the center of
such a galaxy acts as another damper on star formation. A galaxy

merger feeds gas into the black hole, causing it to fire out jets that
heat up gas in the system and prevent it from cooling to form new
stars.

Apparently, once star formation in massive galaxies shuts down,
it does not start up again — most likely because the gas in these systems becomes depleted or becomes so hot that it cannot cool down
quickly enough. These massive galaxies can still merge with one another, but few new stars emerge for want of cold gas. As the massive
galaxies stagnate, smaller galaxies continue to merge and form
stars. The result is that massive galaxies take shape before smaller
ones, as is observed. Dark energy perhaps modulated this process
by determining the degree of galaxy clustering and the rate of
merging.

Dark energy would also explain the evolution of galaxy clusters.
Ancient clusters, found when the universe was less than half its
present age, were already as massive as today's clusters. That is, galaxy clusters have not grown by a significant amount in the past 6
billion to 8 billion years. This lack of growth is an indication that
the infall of galaxies into clusters has been curtailed since the universe was about half its current age — a direct sign that dark energy is influencing the way galaxies are interacting on large scales.
Astronomers knew as early as the mid-1990s that galaxy clusters
had not grown much in the past 8 billion years, and they attributed
this to a lower matter density than theoretical arguments had predicted. The discovery of dark energy resolved the tension between
observation and theory.

An example of how dark energy alters the history of galaxy clusters is the fate of the galaxies in our immediate vicinity, known as
the Local Group. Just a few years ago astronomers thought that the
Milky Way and Andromeda, its closest large neighbor, along with
their retinue of satellites, would fall into the nearby Virgo cluster.
But it now appears that we shall escape that fate and never become
part of a large cluster of galaxies. Dark energy will cause the distance between us and Virgo to expand faster than the Local Group
can cross it.

By throttling cluster development, dark energy also controls the
makeup of galaxies within clusters. The cluster environment fa-
cilitates the formation of a zoo of galaxies, such as the so-called
lenticulars, giant ellipticals, and dwarf ellipticals. By regulating the

ability of galaxies to join clusters, dark energy dictates the relative abundance of these galaxy types.

This is a good story, but is it true? Galaxy mergers, black hole activity, and star formation all decline with time, and very likely they are related in some way. But astronomers have yet to follow the full sequence of events. Ongoing surveys with the Hubble Space Telescope, the Chandra X-ray Observatory, and sensitive ground-based imaging and spectroscopy will scrutinize these links in coming years. One way to do this is to obtain a good census of distant active galaxies and to determine the time when those galaxies last underwent a merger. The analysis will require the development of new theoretical tools but should be within our grasp in the next few years.

Striking a Balance

An accelerating universe dominated by dark energy is a natural way to produce all the observed changes in the galaxy population — namely, the cessation of mergers and its many corollaries, such as loss of vigorous star formation and the end of galactic metamorphosis. If dark energy did not exist, galaxy mergers would probably have continued for longer than they did, and today the universe would contain many more massive galaxies with old stellar populations. Likewise, it would have fewer lower-mass systems, and spiral galaxies such as our Milky Way would be rare (given that spirals cannot survive the merger process). Large-scale structures of galaxies would have been more tightly bound, and more mergers of structures and accretion would have occurred.

Conversely, if dark energy were even stronger than it is, the universe would have had fewer mergers and thus fewer massive galaxies and galaxy clusters. Spiral and low-mass dwarf irregular galaxies would be more common, because fewer galaxy mergers would have occurred throughout time, and galaxy clusters would be much less massive or perhaps not exist at all. It is also likely that fewer stars would have formed, and a higher fraction of our universe's baryonic mass would still be in a gaseous state.

Although these processes may seem distant, the way galaxies form has an influence on our own existence. Stars are needed to produce elements heavier than lithium, which are used to build terrestrial planets and life. If lower star formation rates meant that these elements did not form in great abundance, the universe would not have many planets, and life itself might never have arisen. In this way, dark energy could have had a profound effect on many different and seemingly unrelated aspects of the universe, and perhaps even on the detailed history of our own planet.

Dark energy is by no means finished with its work. It may appear to benefit life: the acceleration will prevent the eventual collapse that was a worry of astronomers not so long ago. But dark energy brings other risks. At the very least, it pulls apart distant galaxies, making them recede so fast that we lose sight of them for good. Space is emptying out, leaving our galaxy and its immediate neighbors an increasingly isolated island. Galaxy clusters, galaxies, and even stars drifting through intergalactic space will eventually have a limited sphere of gravitational influence not much larger than their own individual sizes.

Worse, dark energy might be evolving. Some models predict that if dark energy becomes ever more dominant over time, it will rip apart gravitationally bound objects, such as galaxy clusters and galaxies. Ultimately, planet Earth will be stripped from the sun and shredded, along with all objects on it. Even atoms will be destroyed. Dark energy, once cast in the shadows of matter, will have exacted its final revenge.

GARETH COOK

Untangling the Mystery of the Inca

FROM *Wired*

INCAN CIVILIZATION was a technological marvel. When the Spanish conquistadors arrived in 1532, they found an empire that spanned nearly 3,000 miles, from present-day Ecuador to Chile, all served by a high-altitude road system that included 200-foot suspension bridges built of woven reeds. It was the Inca who constructed Machu Picchu, a cloud city terraced into a precarious stretch of earth hanging between two Andean peaks. They even put together a kind of Bronze Age Internet, a system of messenger posts along the major roads. In one day, Incan runners amped on coca leaves could relay news some 150 miles down the network.

Yet if centuries of scholarship are to be believed, the Inca, whose rule began 2,000 years after Homer, never figured out how to write. It's an enigma known as the Inca paradox, and for nearly 500 years it has stood as one of the great historical puzzles of the Americas. But now a Harvard anthropologist named Gary Urton may be close to untangling the mystery.

His quest revolves around strange, once-colorful bundles of knotted strings called *khipu* (pronounced KEY-poo). The Spanish invaders noticed the khipu soon after arriving but never understood their significance — or how they worked.

Once, at the beginning of the seventeenth century, a group of Spaniards traveling in the central Peruvian highlands east of modern-day Lima encountered an old Indian carrying khipu that he insisted held a record of "all [the Spanish] had done, both the good and the bad." Angered, the Spanish burned the man's khipu, as they did countless others over the years.

Some of the knots did survive, though, and for centuries people wondered if the old man had been speaking the truth. Then in 1923, an anthropologist named Leland Locke provided an answer: the khipu were files. Each knot represented a different number, arranged in a decimal system, and each bundle likely held census data or summarized the contents of storehouses. Roughly a third of the existing khipu don't follow the rules Locke identified, but he speculated that these "anomalous" khipu served some ceremonial or other function. The mystery was considered more or less solved.

Then in the early 1990s, Urton, one of the world's leading Inca scholars, spotted several details that convinced him the khipu contained much more than tallies of llama sales. For example, some knots are tied right over left, others left over right. Urton came to think that this information must signal something. Could the knotted strings also be a form of writing? In 2003 Urton wrote a book outlining his theory, and in 2005 he published a paper in *Science* that showed how even khipu that follow Locke's rules could include place names as well as numbers.

Urton knew that these findings were a tiny part of cracking the code and that he needed the help of people with different skills. So early in 2006, he and a graduate student, Carrie Brezine, unveiled a computerized khipu database — a vast electronic repository that describes every knot on some three hundred khipu in intricate detail. Then Urton and Brezine brought in outside researchers who knew little about anthropology but a lot about mathematics. Led by the Belgian cryptographer Jean-Jacques Quisquater, they are now trying to shake meaning from the knots with a variety of pattern-finding algorithms, one based on a tool used to analyze long strings of DNA, the other similar to Google's PageRank algorithm. They've already identified thousands of repeated knot sequences that suggest words or phrases.

Now the team is closing in on what might be a writing system so unusual that it remained hidden for centuries in plain sight. If successful, the effort will rank with the deciphering of Egyptian hieroglyphics and will let Urton's team rewrite history. But how do you decipher something when it looks completely unlike any known written language — when you're not even sure it has meaning at all?

Urton works a few minutes' walk from Harvard Yard, in a red-brick building with dark wooden doors and copper gutters that also serves as the university's Museum of Natural History. But his fifth-floor office is more Lima than Cambridge. Behind his modest desk hangs a Peruvian pan flute. Spanish-language posters adorn the walls. The space is awash in earthy browns — straw-colored carpet, a darker shade for the faux-clay clock face — set off by colorful weavings hung on every wall. Each object is a memento of his many trips to South America to track down khipu.

Today at least 750 khipu survive, scattered about in museums and private collections. Each one has a long primary cord, typically about a quarter-inch in diameter, from which hang smaller "pendant" cords — sometimes just a couple, sometimes many hundreds. The pendant cords are tied in a series of neat, small knots. Originally dyed in rich colors, the average khipu has now faded so much it resembles a dirty brown mop head.

How could the Inca have used strings to write? In a sense, any written text is just a record of physical actions. You put a pen to paper and then choose from a prescribed set of options how to move and when to lift up. Each decision is preserved in ink. The same can be done with string. The writer makes a series of decisions, recorded as a knot that can then be read by anyone who knows the rules.

Back in the 1920s, Locke began with the observation that the Inca tied their khipu with three types of knots. There is a figure-eight knot, which represents one of something. There are long knots, with two to nine turns, representing those numbers. And there are single knots, which represent tens, hundreds, thousands, or ten thousands, depending on where they fall on the string. When a khipu is placed flat on the ground, the bottom row is the ones place, and successively higher rows stand for higher places. So the number 327 would have three single knots in the hundreds place. A little lower would be two single knots. Lower still would be a long knot with seven turns.

Most anthropologists assumed that was all there was to it — until 1992. That's when Urton spent a day looking at khipu in the American Museum of Natural History in New York with his friend Bill Conklin, an architect and textile expert. As he studied the cords, Conklin had an isn't-that-funny insight: the knots that connect the small pendant strings to the primary cord are always tied the same way, but sometimes they face forward and sometimes backward. Startled, Urton soon noticed additional construction details — such as whether a fiber had been dyed to have a bluish or a reddish tint. All told, Urton has found seven additional bits of binary information that might signal something. Perhaps one means "read this as a word, not a number." Perhaps the binary code served as a kind of markup language, allowing the Inca to make notes on top of Locke's number-recording system. And perhaps the two hundred or so anomalous khipu don't follow Locke's rules because they've transcended them.

Most Incan scholars are intrigued by Urton's ideas, though a few skeptics have noted that he has not produced any proof that his binary code carries meaning, much less that the khipu contain narratives. The Harvard professor concedes that some of the information he's looking at may not signal anything. But he is convinced the khipu have stories to tell, and he has some history on his side. José de Acosta, a Jesuit missionary sometimes called the Pliny of the New World, wrote a description of the khipu at the end of the sixteenth century. In it he described how the "woven reckonings" were used to record financial transactions involving hens, eggs, and hay. But he also noted that the native people considered the khipu to be "witnesses and authentic writing." "I saw a bundle of these strings," he wrote, "on which a woman had brought a written confession of her whole life and used it to confess just as I would have done with words written on paper."

Egyptian hieroglyphics, Linear B, ancient Mayan writing — all of the great decipherments have been accomplished by a combination of logic and intuition, persistence and flexibility. Decoding scripts is not like looking for a combination that will open a lock. It's more like rock climbing: you find a foothold, push up, and hope that another presents itself.

Jean-Jacques Quisquater — a tall man with a thin crown of wispy white hair — would like to join the pantheon of puzzle solvers. Quisquater directs a large cryptography laboratory at Belgium's historic Catholic University of Louvain, where he is known for his work on securing smart cards. In the fall of 2003, he came to MIT for a yearlong academic sabbatical. At the time, he had been thinking nostalgically of a trip to Greece forty years before, when he saw the famous undeciphered Phaistos Disk, a small red-brown disk from deep in the second millennium B.C. covered on each side with a spiral of glyphs: a fish, a shield, an olive branch. Quisquater hoped to find something equally romantic and challenging to work on.

When he heard about the mystery of the khipu, he was immediately enthralled. He soon met Urton, and they teamed up with a father-son pair of MIT computer scientists, Martin and Erik Demaine. The group began chatting, with the mathematicians offering detailed plans about how to sort the data.

The team agreed that one of Quisquater's graduate students, Vincent Castus, would first try an analysis known as a suffix tree. The method uses a computer to identify all the blocks of characters in a text that repeat themselves. Thus the word "Mississippi" would yield several repeated blocks, including *issi, iss,* and *ss.* Suffix trees are used in genetic analysis to find the shortest unique pattern in a sample of DNA.

With the khipu database loaded onto his iMac, Castus worked to build a suffix tree from the knots, leaving aside the more complicated binary data on this first pass. He began in May 2006. By October he had worked out all the details and found an astonishing number of repeats: 3,000 different groups of repeated five-knot sequences. Shorter patterns appeared even more often. He found several pairs of khipu linked by large numbers of matches, suggesting that they could be related.

None of this tells us whether the khipu contain words or stories. It's possible the researchers have found khipu that just happen to include repeated number sequences that are not interesting for any particular reason or that some khipu are deliberate copies of others.

But Urton suspects there's more to it than that. He knows repetition is the code-breaker's great friend. A Cold War sleuth noticing an oft-used sequence might guess it stood for Moscow or Khrushchev. Recognizing repeated place names was one of the first steps in deciphering the ancient Mycenaean script Linear B. Now the team has a key for all the khipu in the database, allowing them to instantly identify whenever a particular sequence appears. They also have a list of common short sequences — the most obvious candidates for words.

The team had previously made one breakthrough in identifying connections between knots, thanks to Brezine, who has a background in mathematics and just happens to be a weaver on the side. The master of the khipu database, she wanted to find examples of strings with numbers that added up to sums on another khipu. So she developed a simple algorithm and combed through the data.

Her efforts identified a handful of interlinked khipu that had been uncovered together in a cache in Puruchuco, an archaeological site near Lima. The khipu looked like records kept by three successively higher levels of Incan administrators. Add the numbers on one khipu, and the sum is found on another, with that sum in turn found on a third. Imagine, for example, that they depict the results of a census. The village counts up its people and then forwards the total to the district. The district records the numbers from several villages and then forwards the results up to the provincial head. Urton and Brezine do not know what is being counted (people? llamas?), but their 2005 *Science* paper showed for the first time that information flowed between the khipu.

They have also identified what may be the first word. The two higher-level khipu in the census example use an introductory sequence of three figure-eight knots (1-1-1) that does not appear on what they assume is the village-level khipu. Perhaps only the upper layers have the sequence because it is a label for a particular place, used when compiling information from many locations. Maybe, they suggest, the first symbol to be read off a khipu means this: Puruchuco.

Quisquater's team, meanwhile, is now working on another, even more ambitious way of extracting clues. It depends on thinking of each knot as a node and each khipu as a network and the links being lengths of string.

One of the surprises from the burgeoning new field of network theory is that the role of a particular node can be summarized — in a deep and meaningful way — by a single number. A good example of this is Google's PageRank algorithm. The power of the company's search engine comes from its ability to rank Web pages by relevance. On the Web, a link runs from one page to another like an arrow. The algorithm interprets that as the first page voting for the second one. Votes flow from across the Internet, like streams joining rivers, eventually pooling at the eBays of the world.

The analysis that the team plans for these khipu networks doesn't exactly mimic PageRank. After all, the string links between knots aren't unidirectional like arrows; one knot doesn't point to another. But the concept is the same: if you think about a big mass of information as a network and analyze it as a network, looking for the thousands of small and big ways that different piles of information relate to one another, you can see things that you wouldn't notice otherwise.

Vincent Blondel, a Belgian mathematics professor who is a friend of Quisquater's, recently helped work out the math behind an approach that allows a computer to calculate degrees of similarity between nodes in two separate networks. Like PageRank, the procedure uses voting, but it assigns each node many scores instead of one and employs a more complex scheme for calculating the totals. Type "baseball" into Google, and its spiders will race over the Internet, look at links, and spit back that yankees.com is the eleventh most useful site for you and seattlemariners.com is the twenty-second. If Quisquater's algorithm were used on the Web, it would return a slew of numbers, some of which would show similarities between different nodes — or knots. So you'd see that the Yankees and Mariners sites are similar because both receive feeds from majorleaguebaseball.com and have outgoing links to the home pages for twenty-nine teams.

When Quisquater's algorithm is used on khipu, it will reveal knots or groups of knots that always play a certain role in relationship to others. These might be labels or formatting signs. For example, it may turn out that some of the khipu start with sets of knots that say something like "read this as a calendar." Or collections of khipu may have similar networks of closely related knots, perhaps signaling that they originate from the same geographic area. Or it could even turn out that the anomalous khipu will all have some pattern that signifies "read this as a story." The results from this technique were expected to come in sometime in 2007, and they will provide valuable clues, even if they don't immediately crack the Inca paradox.

*

Urton's great insight has been to treat the khipu not just as a textile or a simple abacus but as an advanced alien technology. Sitting on a poncho draped over the couch in his office, Urton describes a formative trip to a remote Bolivian village where he worked with traditional weavers. Observing these women spin and weave yarn into multicolored tapestries with elaborate symmetries, he caught a glimpse of the Incan mind at work. For an expert weaver, fabric is a record of many choices, a dance of twists, turns, and pulls that leads to the final product. They would have seen a fabric — be it cloth or knotted strings — a bit the way a chess master views a game in progress. Yes, they see a pattern of pieces on a board, but they also have a feel for the moves that led there.

"You can see inside of it," Urton says.

It would be all too easy to dismiss the khipu as the work of a less advanced civilization, one that didn't develop guns, iron, or wheels. But for more than a decade, Urton has assumed that the khipu are evidence of Incan sophistication in ways we have still not grasped.

Acosta, the sixteenth-century Jesuit missionary, believed this. He traveled throughout the Americas and recorded several observations of khipu in use. He described religious converts memorizing prayers using khipu-like devices made of small stones or kernels of corn. He also described people in a churchyard completing difficult calculations "without making the slightest error . . . Whoever wants may judge whether this is clever or if these people are brutish," he wrote, "but I judge it is certain that, in that which they here apply themselves, they get the better of us."

C. JOSH DONLAN

Restoring America's Big, Wild Animals

FROM *Scientific American*

IN THE FALL OF 2004 a dozen conservation biologists gathered on a ranch in New Mexico to ponder a bold plan. The scientists, trained in a variety of disciplines, ranged from the grand old men of the field to those of us earlier in our careers. The idea we were mulling over was the reintroduction of large vertebrates — megafauna — to North America. Most of these animals, such as mammoths and cheetahs, died out roughly 13,000 years ago, when humans from Eurasia began migrating to the continent. The theory — propounded forty years ago by Paul Martin of the University of Arizona — is that overhunting by the new arrivals reduced the numbers of large vertebrates so severely that the populations could not recover. Called Pleistocene overkill, the concept was highly controversial at the time, but the general thesis that humans played a significant role is now widely accepted. Martin was present at the meeting in New Mexico, and his ideas on the loss of these animals, the ecological consequences, and what we should do about it formed the foundation of the proposal that emerged, which we dubbed Pleistocene rewilding.

Although the cheetahs, lions, and mammoths that once roamed North America are extinct, the same species or close relatives have survived elsewhere, and our discussions focused on introducing these substitutes to North American ecosystems. We believe that these efforts hold the potential to partially restore important ecological processes, such as predation and browsing, to ecosystems in which they have been absent for millennia. The substitutes would also bring economic and cultural benefits. Not surprisingly, the published proposal evoked strong reactions. Those reactions are welcome, because debate about the conservation issues that underlie Pleistocene rewilding merit thorough discussion.

Why Big Animals Are Important

Our approach concentrates on large animals because they exercise a disproportionate effect on the environment. For tens of millions of years, megafauna dominated the globe, strongly interacting and coevolving with other species and influencing entire ecosystems. Horses, camels, lions, elephants, and other large creatures were everywhere: megafauna were the norm. But starting roughly 50,000 years ago, the overwhelming majority went extinct. Today megafauna inhabit less than 10 percent of the globe.

Over the past decade, the ecologist John Terborgh of Duke University has observed directly how critical large animals are to the health of ecosystems and how their loss adversely affects the natural world. When a hydroelectric dam flooded thousands of acres in Venezuela, Terborgh saw the water create dozens of islands — a fragmentation akin to the virtual islands created around the world as humans cut down trees, build shopping malls, and sprawl from urban centers. The islands in Venezuela were too small to support the creatures at the top of the food chain — predators such as jaguars, pumas, and eagles. Their disappearance sparked a chain of reactions. Animals such as monkeys, leafcutter ants, and other herbivores, whose populations were no longer kept in check by predation, thrived and subsequently destroyed vegetation; the ecosystems collapsed, with biodiversity being the ultimate loser.

Similar ecological disasters have occurred on other continents. Degraded ecosystems are not only bad for biodiversity; they are bad for human economies. In Central America, for instance, researchers have shown that intact tropical ecosystems are worth at least $60,000 a year to a single coffee farm because of the services they provide, such as pollination of coffee crops.

Where large predators and herbivores still remain, they play pivotal roles. In Alaska, sea otters maintain kelp-forest ecosystems by keeping herbivores that eat kelp, such as sea urchins, in check. In Africa, elephants are keystone players; as they move through an area, knocking down trees and trampling vegetation, they create a habitat in which certain plants and animals can flourish. Lions and other predators control the populations of African herbivores, which in turn influence the distribution of plants and soil nutrients.

In Pleistocene America, large predators and herbivores played similar roles. Today most of that vital influence is absent. For example, the American cheetah (a relative of the African cheetah) dashed across the grasslands in pursuit of pronghorn antelopes for millions of years. These chases shaped the pronghorn's astounding speed and other biological aspects of one of the fastest animals alive. In the absence of the cheetah, the pronghorn appears "overbuilt" for its environment today.

Pleistocene rewilding is not about recreating exactly some past state. Rather it is about restoring the kinds of species interactions that sustain thriving ecosystems. Giant tortoises, horses, camels, cheetahs, elephants, and lions — they were all here, and they helped shape North American ecosystems. Either the same species or closely related species are available for introduction as proxies, and many are already in captivity in the United States. In essence, Pleistocene rewilding would help change the underlying premise of conservation biology from limiting extinction to actively restoring natural processes.

At first, our proposal may seem outrageous — lions in Montana? But the plan deserves serious debate for several reasons. First, no place on Earth is pristine, at least in terms of being substantially free of human influence. Our demographics, chemicals, economics, and politics pervade every part of the planet. Even in our largest national parks, species go extinct without active intervention. And human encroachment shows alarming signs of worsening. Bold actions, rather than business as usual, will be needed to reverse such negative influences. Second, ever since conservation biology emerged as a discipline more than three decades ago, it has been mainly a business of doom and gloom, a struggle merely to slow the loss of biodiversity. But conservation need not be only reactive. A proactive approach would include restoring natural processes, starting with ones we know are disproportionately important, such as those influenced by megafauna.

Third, land in North America is available for the reintroduction of megafauna. Although the patterns of human land use are always shifting, in some areas, such as parts of the Great Plains and the Southwest, large private and public lands with low or declining human population densities might be used for the project. Fourth, bringing megafauna back to America would also bring tourist dollars and other income to nearby communities and enhance the public's appreciation of the natural world. More than 1.5 million people visit San Diego's Wild Animal Park every year to catch a glimpse of large mammals. Only a handful of American national parks receive that many visitors. Last, the loss of some of the remaining species of megafauna in Africa and Asia within this century seems likely, and Pleistocene rewilding could help reverse that.

How It Might Be Done

We are not talking about backing up a van and kicking some cheetahs out into your backyard. Nor are we talking about doing it tomorrow. We conceive of Pleistocene rewilding as a series of staged, carefully managed ecosystem manipulations. What we are offering here is a vision — not a blueprint — of how this might be accomplished. And by no means are we suggesting that rewilding should be a priority over current conservation programs in North America or Africa. Pleistocene rewilding could proceed alongside such conservation efforts, and it would likely generate conservation dollars from new funding sources rather than competing for funds with existing conservation efforts.

The long-term vision includes a vast, securely fenced ecological history park, encompassing thousands of square miles, where horses, camels, elephants, and large carnivores would roam. As happens now in Africa and regions surrounding some U.S. national parks, the ecological history park would not only attract ecotourists but would also provide jobs related both to park management and to tourism.

To get to that distant point, we would need to start modestly, with relatively small-scale experiments that assess the impacts of megafauna on North American landscapes. These controlled experiments, guided by sound science and by the fossil record, which indicates what animals actually lived here, could occur first on donated or purchased private lands and could begin immediately.

They will be critical in answering the many questions about the reintroductions and would help lay out the costs and benefits of rewilding.

One of these experiments is already under way. Spurred by our 2004 meeting, biologists recently reintroduced Bolson tortoises to a private ranch in New Mexico. Bolson tortoises, some weighing more than 100 pounds, once grazed parts of the southwestern United States before disappearing around 10,000 years ago, victims of human hunting. This endangered tortoise now clings to survival, restricted to a single small area in central Mexico. Thus the reintroduction not only repatriates the tortoise to the United States, it increases the species' chance for survival. Similar experiments are also occurring outside North America.

The reintroduction of wild horses and camels would be a logical part of these early experiments. Horses and camels originated on this continent, and many species were present in the late Pleistocene. Today's feral horses and asses that live in some areas of the West are plausible substitutes for extinct American species. Because most of the surviving Eurasian and African species are now critically endangered (see "Endangered Wild Equids," by Patricia D. Moehlman, *Scientific American,* March 2005), establishing Asian asses and Przewalski's horse in North America might help prevent the extinction of these animals. Bactrian camels, which are critically endangered in the Gobi Desert, could provide a modern proxy for *Camelops,* a late Pleistocene camel. Camels, introduced from captive or domesticated populations, might benefit U.S. ecosystems by browsing on woody plants that today are overtaking arid grasslands in the Southwest, an ecosystem that is increasingly endangered.

Another prong of the project would likely be more controversial but could also begin immediately. It would establish small numbers of elephants, cheetahs, and lions on private property.

Introducing elephants could prove valuable to nearby human populations by attracting tourists and maintaining grasslands useful to ranchers (elephants could suppress the woody plants that threaten southwestern grasslands). In the late Pleistocene, at least four elephant species lived in North America. Under a scientific framework, captive elephants in the United States could be introduced as proxies for these extinct animals. The biggest cost involved would be fencing, which has helped reduce conflict between elephants and humans in Africa.

Many cheetahs are already in captivity in the United States. The greatest challenge would be to provide them with large, securely fenced areas that have appropriate habitat and prey animals. Offsetting these costs are benefits: restoring what must have been strong interactions with pronghorns, facilitating ecotourism as an economic alternative for ranchers, many of whom are struggling financially, and helping to save the world's fastest carnivore from extinction.

Lions are increasingly threatened, with populations in Asia and some parts of Africa critically endangered. Bringing back lions, which are the same species that once lived in North America, presents daunting challenges as well as many potential benefits. But private reserves in southern Africa where lions and other large animals have been successfully reintroduced offer a model — and these reserves are smaller than some private ranches in the Southwest.

If these early experiments with large herbivores and predators show promising results, more could be undertaken, moving toward the long-term goal of a huge ecological history park. What we need now are panels of experts who, for each species, could assess, advise, and cautiously lead efforts in restoring megafauna to North America.

A real-world example of how the reintroduction of a top predator might work comes from the wolves of Yellowstone National Park (see "Lessons from the Wolf," by Jim Robbins, *Scientific American,* June 2004). The gray wolf became extinct in and around Yellowstone during the 1920s. The loss led to increases in their prey — moose and elk — which in turn reduced the distribution of aspens and other trees they eat. Lack of vegetation destroyed habitat for migratory birds and beavers. Thus the disappearance of the wolves propagated a trophic cascade from predators to herbivores to plants to birds and beavers. Scientists have started to document the ecosystem changes as reintroduced wolves regain the ecological role they played in Yellowstone for millennia. An additional insight researchers are learning from putting wolves back into Yellowstone is that they may be helping the park cope with climate change. As winters grow milder, fewer elk die, which means less carrion for scavengers such as coyotes, ravens, and bald eagles. Wolves provide carcasses throughout the winter for the scavengers to feed on, bestowing a certain degree of stability.

The Challenges Ahead

As our group on the ranch in New Mexico discussed how Pleistocene rewilding might work, we foresaw many challenges that would have to be addressed and overcome. These include the possibility that introduced animals could bring novel diseases with them or that they might be unusually susceptible to diseases already present in the ecosystem; the fact that habitats have changed over the millennia and that reintroduced animals might not fare well in these altered environments; and the likelihood of unanticipated ecological consequences and unexpected reactions from neighboring human communities. Establishing programs that monitor the interactions among species and their consequences for the well-being of the ecosystem will require patience and expertise. And, of course, it will not be easy to convince the public to accept predation as an important natural process that actually nourishes the land and enables ecosystems to thrive. Other colleagues have raised additional concerns, albeit none that seem fatal.

Many people will claim that the concept of Pleistocene rewilding is simply not feasible in the world we live in today. I urge these people to look to Africa for inspiration. The year after the creation of Kruger National Park was announced, the site was hardly the celebrated mainstay of southern African biodiversity it is today. In 1903 zero elephants, nine lions, eight buffalo, and very few cheetahs lived within its boundaries. Thanks to the vision and dedication of African conservationists, 7,300 elephants, 2,300 lions, 28,000 buffalo, and 250 cheetahs roamed Kruger a hundred years later — as did 700,000 tourists, bringing with them tens of millions of dollars.

In the coming century, humanity will decide, by default or design, the extent to which it will tolerate other species and thus how much biodiversity will endure. Pleistocene rewilding is not about trying to go back to the past; it is about using the past to inform society about how to maintain the functional fabric of nature. The potential scientific, conservational, and cultural benefits of restoring megafauna are clear, as are the costs. Although sound science can help mitigate the potential costs, these ideas will make many uneasy. Yet given the apparent dysfunction of North American ecosystems and Earth's overall state, inaction carries risks as well. In the face of tremendous uncertainty, science and society must weigh the costs and benefits of bold, aggressive actions like Pleistocene rewilding against those of business as usual, which has risks, uncertainties, and costs that are often unacknowledged. We have a tendency to think that if we maintain the status quo, things will be fine. All the available information suggests the opposite.

FREEMAN DYSON

Our Biotech Future

FROM *New York Review of Books*

IT HAS BECOME PART of the accepted wisdom to say that the twentieth century was the century of physics and the twenty-first century will be the century of biology. Two facts about the coming century are agreed on by almost everyone. Biology is now bigger than physics, as measured by the size of budgets, the size of the workforce, or the output of major discoveries; and biology is likely to remain the biggest part of science through the twenty-first century. Biology is also more important than physics, as measured by its economic consequences, its ethical implications, or its effects on human welfare.

These facts raise an interesting question. Will the domestication of high technology, which we have seen marching from triumph to triumph with the advent of personal computers and GPS receivers and digital cameras, soon be extended from physical technology to biotechnology? I believe that the answer to this question is yes. Here I am bold enough to make a definite prediction. I predict that the domestication of biotechnology will dominate our lives during the next fifty years at least as much as the domestication of computers has dominated our lives during the past fifty years.

I see a close analogy between John von Neumann's blinkered vision of computers as large centralized facilities and the public perception of genetic engineering today as an activity of large pharmaceutical and agribusiness corporations such as Monsanto. The public distrusts Monsanto because Monsanto likes to put genes for poisonous pesticides into food crops, just as we distrusted von Neumann because he liked to use his computer for designing hydrogen bombs secretly at midnight. It is likely that genetic engineering will remain unpopular and controversial so long as it remains a centralized activity in the hands of large corporations.

I see a bright future for the biotechnology industry when it follows the path of the computer industry, the path that von Neumann failed to foresee, becoming small and domesticated rather than big and centralized. The first step in this direction was taken recently when genetically modified tropical fish with new and brilliant colors appeared in pet stores. For biotechnology to become domesticated, the next step is to become user-friendly. I recently spent a happy day at the Philadelphia Flower Show, the biggest indoor flower show in the world, where flower breeders from all over the world show off the results of their efforts. I have also visited the reptile show in San Diego, an equally impressive show displaying the work of another set of breeders. Philadelphia excels in orchids and roses, San Diego excels in lizards and snakes. The main problem for a grandparent visiting the reptile show with a grandchild is to get the grandchild out of the building without actually buying a snake.

Every orchid or rose or lizard or snake is the work of a dedicated and skilled breeder. There are thousands of people, amateurs and professionals, who devote their lives to this business. Now imagine what will happen when the tools of genetic engineering become accessible to these people. There will be do-it-yourself kits for gardeners, who will use genetic engineering to breed new varieties of roses and orchids. Also kits for lovers of pigeons and parrots and lizards and snakes to breed new varieties of pets. Breeders of dogs and cats will have their kits too.

Domesticated biotechnology, once it gets into the hands of housewives and children, will give us an explosion of diversity of new living creatures rather than the monoculture crops that the big corporations prefer. New lineages will proliferate to replace those that monoculture farming and deforestation have destroyed. Designing genomes will be a personal thing, a new art form as creative as painting or sculpture.

Few of the new creations will be masterpieces, but a great many will bring joy to their creators and variety to our fauna and flora. The final step in the domestication of biotechnology will be biotech games, designed like computer games for children down to kindergarten age but played with real eggs and seeds rather than with images on a screen. Playing such games, kids will acquire an intimate feeling for the organisms that they are growing. The winner could be the kid whose seed grows the prickliest cactus or the kid whose egg hatches the cutest dinosaur. These games will be messy and possibly dangerous. Rules and regulations will be needed to make sure that our kids do not endanger themselves and others. The dangers of biotechnology are real and serious.

If domestication of biotechnology is the wave of the future, five important questions need to be answered. First, can it be stopped? Second, ought it to be stopped? Third, if stopping it is either impossible or undesirable, what are the appropriate limits that our society must impose on it? Fourth, how should the limits be decided? Fifth, how should the limits be enforced, nationally and internationally? I do not attempt to answer these questions here. I leave it to our children and grandchildren to supply the answers.

A New Biology for a New Century

Carl Woese is the world's greatest expert in the field of microbial taxonomy, the classification and understanding of microbes. He explored the ancestry of microbes by tracing the similarities and differences between their genomes. He discovered the large-scale structure of the tree of life, with all living creatures descended from three primordial branches. Before Woese, the tree of life had two main branches, called prokaryotes and eukaryotes, the prokaryotes composed of cells without nuclei and the eukaryotes composed of cells with nuclei. All kinds of plants and animals, including humans, belonged to the eukaryote branch. The prokaryote branch contained only microbes. Woese discovered, by studying the anatomy of microbes in detail, that there are two fundamentally different kinds of prokaryotes, which he called bacteria and archea. So he constructed a new tree of life with three branches: bacteria, archea, and eukaryotes. Most of the well-known microbes are bacteria. The archea were at first supposed to be rare and confined to extreme environments such as hot springs, but they are now known to be abundant and widely distributed over the planet. Woese recently published two provocative and illuminating articles with the titles "A New Biology for a New Century" and, with Nigel Goldenfeld, "Biology's Next Revolution."

Woese's main theme is the obsolescence of reductionist biology as it has been practiced for the last hundred years, with its assumption that biological processes can be understood by studying genes and molecules. What is needed instead is a new synthetic biology based on emergent patterns of organization. Aside from his main theme, he raises another important question. When did Darwinian evolution begin? By Darwinian evolution he means evolution as Darwin understood it, based on the competition for survival of noninterbreeding species. He presents evidence that Darwinian evolution does not go back to the beginning of life. When we compare genomes of ancient lineages of living creatures, we find evidence of numerous transfers of genetic information from one lineage to another. In early times, horizontal gene transfer, the sharing of genes between unrelated species, was prevalent. It becomes more prevalent the further back you go in time.

Whatever Carl Woese writes, even in a speculative vein, needs to be taken seriously. In his "New Biology" article, he is postulating a golden age of pre-Darwinian life, when horizontal gene transfer was universal and separate species did not yet exist. Life was then a community of cells of various kinds, sharing their genetic information so that clever chemical tricks and catalytic processes invented by one creature could be inherited by all of them. Evolution was a communal affair, the whole community advancing in metabolic and reproductive efficiency as the genes of the most efficient cells were shared. Evolution could be rapid, as new chemical devices could be evolved simultaneously by cells of different kinds working in parallel and then reassembled in a single cell by horizontal gene transfer.

But then one evil day, a cell resembling a primitive bacterium happened to find itself one jump ahead of its neighbors in efficiency. That cell, anticipating Bill Gates by 3 billion years, separated itself from the community and refused to share. Its offspring became the first species of bacteria — and the first species of any kind — to reserve their intellectual property for their own private use. With their superior efficiency, the bacteria continued to prosper and to evolve separately, while the rest of the community continued its communal life. Some millions of years later, another cell separated itself from the community and became the ancestor of the archea. Sometime after that, a third cell separated itself and became the ancestor of the eukaryotes. And so it went on, until nothing was left of the community and all life was divided into species. The Darwinian interlude had begun.

The Darwinian interlude has lasted for 2 or 3 billion years. It probably slowed down the pace of evolution considerably. The basic biochemical machinery of life had evolved rapidly during the few hundreds of millions of years of the pre-Darwinian era, and it changed very little in the next 2 billion years of microbial evolution. Darwinian evolution is slow because individual species, once established, evolve very little. With rare exceptions, Darwinian evolution requires established species to become extinct so that new species can replace them.

Now, after 3 billion years, the Darwinian interlude is over. It was an interlude between two periods of horizontal gene transfer. The epoch of Darwinian evolution based on competition between species ended about 10,000 years ago, when a single species, *Homo sapiens,* began to dominate and reorganize the biosphere. Since that time, cultural evolution has replaced biological evolution as the main driving force of change. Cultural evolution is not Darwinian. Cultures spread by horizontal transfer of ideas more than by genetic inheritance. Cultural evolution is running a thousand times faster than Darwinian evolution, taking us into a new era of cultural interdependence, which we call globalization. And now, as *Homo sapiens* domesticates the new biotechnology, we are reviving the ancient pre-Darwinian practice of horizontal gene transfer, moving genes easily from microbes to plants and animals, blurring the boundaries between species. We are moving rapidly into the post-Darwinian era, when species other than our own will no longer exist, and the rules of open-source sharing will be extended from the exchange of software to the exchange of genes. Then the evolution of life will once again be communal, as it was in the good old days before separate species and intellectual property were invented.

I would like to borrow Carl Woese's vision of the future of biology and extend it to the whole of science. Here is his metaphor for the future of science:

Imagine a child playing in a woodland stream, poking a stick into an eddy in the flowing current, thereby disrupting it. But the eddy quickly reforms. The child disperses it again. Again it reforms, and the fascinating game goes on. There you have it! Organisms are resilient patterns in a turbulent flow — patterns in an energy flow ... It is becoming increasingly clear that to understand living systems in any deep sense, we must come to see them not materialistically, as machines, but as stable, complex, dynamic organization.

This picture of living creatures as patterns of organization rather than collections of molecules applies not only to bees and bacteria, butterflies and rain forests, but also to sand dunes and snowflakes, thunderstorms and hurricanes. The nonliving universe is as diverse and dynamic as the living universe and is also dominated by patterns of organization that are not yet understood. The reductionist physics and reductionist molecular biology of the twentieth century will continue to be important in the twenty-first century, but they will not be dominant. The big problems, the evolution of the universe as a whole, the origin of life, the nature of human consciousness, and the evolution of Earth's climate, cannot be understood by reducing them to elementary particles and molecules. New ways of thinking and new ways of organizing large databases will be needed.

Green Technology

The domestication of biotechnology in everyday life may also be helpful in solving practical economic and environmental problems. Once a new generation of children has grown up, as familiar with biotech games as our grandchildren are now with computer games, biotechnology will no longer seem weird and alien. In the era of open-source biology, the magic of genes will be available to anyone with the skill and imagination to use it. The way will be open for biotechnology to move into the mainstream of economic development, to help us solve some of our urgent social problems and ameliorate the human condition all over Earth. Open-source biology could be a powerful tool, giving us access to cheap and abundant solar energy.

A plant is a creature that uses the energy of sunlight to convert water and carbon dioxide and other simple chemicals into roots and leaves and flowers. To live, it needs to collect sunlight. But it uses sunlight with low efficiency. The most efficient crop plants, such as sugar cane or maize, convert about 1 percent of the sunlight that falls onto them into chemical energy. Artificial solar collectors made of silicon can do much better. Silicon solar cells can convert sunlight into electrical energy with 15 percent efficiency, and electrical energy can be converted into chemical energy without much loss. We can imagine that in the future, when we have mastered the art of genetically engineering plants, we may breed new crop plants that have leaves made of silicon, converting sunlight into chemical energy with ten times the efficiency of natural plants. These artificial crop plants would reduce the area of land needed for biomass production by a factor of ten. They would allow solar energy to be used on a massive scale without taking up too much land. They would look like natural plants except that their leaves would be black, the color of silicon, instead of green, the color of chlorophyll. The question I am asking is, how long will it take us to grow plants with silicon leaves?

If the natural evolution of plants had been driven by the need for high efficiency of utilization of sunlight, then the leaves of all plants would have been black. Black leaves would absorb sunlight more efficiently than leaves of any other color. Obviously plant evolution was driven by other needs, and in particular by the need for protection against overheating. For a plant growing in a hot climate, it is advantageous to reflect as much as possible of the sunlight that is not used for growth. There is plenty of sunlight, and it is not important to use it with maximum efficiency. The plants have evolved with chlorophyll in their leaves to absorb the useful red and blue components of sunlight and to reflect the green. That is why it is reasonable for plants in tropical climates to be green. But this logic does not explain why plants in cold climates where sunlight is scarce are also green. We could imagine that in a place like Iceland, overheating would not be a problem, and plants with black leaves using sunlight more efficiently would have an evolutionary advantage. For some reason that we do not understand, natural plants with black leaves never appeared. Why not? Perhaps we shall not understand why nature did not travel this route until we have traveled it ourselves.

After we have explored this route to the end, when we have created new forests of black-leaved plants that can use sunlight ten times more efficiently than natural plants, we shall be confronted by a new set of environmental problems. Who shall be allowed to grow the black-leaved plants? Will black-leaved plants remain an artificially maintained cultivar, or will they invade and permanently change the natural ecology? What shall we do with the silicon trash that these plants leave behind them? Shall we be able to design a whole ecology of silicon-eating microbes and fungi and earthworms to keep the black-leaved plants in balance with the rest of nature and to recycle their silicon? The twenty-first century will bring us powerful new tools of genetic engineering with which to manipulate our farms and forests. With the new tools will come new questions and new responsibilities.

Rural poverty is one of the great evils of the modern world. The lack of jobs and economic opportunities in villages drives millions of people to migrate from villages into overcrowded cities. The continuing migration causes immense social and environmental problems in the major cities of poor countries. The effects of poverty are most visible in the cities, but the causes of poverty lie mostly in the villages. What the world needs is a technology that directly attacks the problem of rural poverty by creating wealth and jobs in the villages. A technology that creates industries and careers in villages would give the villagers a practical alternative to migration. It would give them a chance to survive and prosper without uprooting themselves.

The shifting balance of wealth and population between villages and cities is one of the main themes of human history over the last 10,000 years. The shift from villages to cities is strongly coupled with a shift from one kind of technology to another. I find it convenient to call the two kinds of technology green and gray. The adjective "green" has been appropriated and abused by various political movements, especially in Europe, so I need to explain clearly what I have in mind when I speak of green and gray. Green technology is based on biology; gray technology, on physics and chemistry.

Roughly speaking, green technology is the technology that gave birth to village communities 10,000 years ago, starting with the domestication of plants and animals, the invention of agriculture, the breeding of goats and sheep and horses and cows and pigs, the manufacture of textiles and cheese and wine. Gray technology is the technology that gave birth to cities and empires 5,000 years later, starting with the forging of bronze and iron, the invention of wheeled vehicles and paved roads, the building of ships and war chariots, the manufacture of swords and guns and bombs. Gray technology also produced the steel plows, tractors, reapers, and processing plants that made agriculture more productive and transferred much of the resulting wealth from village-based farmers to city-based corporations.

For the first 5,000 of the 10,000 years of human civilization, wealth and power belonged to villages with green technology, and for the second 5,000 years wealth and power belonged to cities with gray technology. Beginning about 500 years ago, gray technology became increasingly dominant, as we learned to build machines that used power from wind and water and steam and electricity. In the last hundred years, wealth and power were even more heavily concentrated in cities as gray technology raced ahead. As cities became richer, rural poverty deepened.

This sketch of the last 10,000 years of human history puts the problem of rural poverty in a new perspective. If rural poverty is a consequence of the unbalanced growth of gray technology, it is possible that a shift in the balance back from gray to green might cause rural poverty to disappear. That is my dream. During the last fifty years we have seen explosive progress in the scientific understanding of the basic processes of life, and in the last twenty years this new understanding has given rise to explosive growth of green technology. The new green technology allows us to breed new varieties of animals and plants as our ancestors did 10,000 years ago, but now a hundred times faster. It now takes us a decade instead of a millennium to create new crop plants, such as the herbicide-resistant varieties of maize and soybeans that allow weeds to be controlled without plowing and greatly reduce the erosion of topsoil by wind and rain. Guided by a precise understanding of genes and genomes instead of by trial and error, we can within a few years modify plants so as to give them improved yield, improved nutritive value, and improved resistance to pests and diseases.

Within a few more decades, as the continued exploring of genomes gives us better knowledge of the architecture of living creatures, we shall be able to design new species of microbes and plants according to our needs. The way will then be open for green technology to do more cheaply and cleanly many of the things that gray technology can do, and also to do many things that gray technology has failed to do. Green technology could replace most of our existing chemical industries and a large part of our mining and manufacturing industries. Genetically engineered earthworms could extract common metals such as aluminum and titanium from clay, and genetically engineered seaweed could extract magnesium or gold from seawater. Green technology could also achieve more extensive recycling of waste products and worn-out machines, with great benefit to the environment. An economic system based on green technology could come much closer to the goal of sustainability, using sunlight instead of fossil fuels as the primary source of energy. New species of termites could be engineered to chew up derelict automobiles instead of houses, and new species of trees could be engineered to convert carbon dioxide and sunlight into liquid fuels instead of cellulose.

Before genetically modified termites and trees can be allowed to help solve our economic and environmental problems, great arguments will rage over the possible damage they may do. Many of the people who call themselves green are passionately opposed to green technology. But in the end, if the technology is developed carefully and deployed with sensitivity to human feelings, it is likely to be accepted by most of the people who will be affected by it, just as the equally unnatural and unfamiliar green technologies of milking cows and plowing soils and fermenting grapes were accepted by our ancestors long ago. I am not saying that the political acceptance of green technology will be quick or easy. I say only that green technology has enormous promise for preserving the balance of nature on this planet as well as for relieving human misery. Future generations of people raised from childhood with biotech toys and games will probably accept it more easily than we do. Nobody can predict how long it may take to try out the new technology in a thousand different ways and measure its costs and benefits.

What has this dream of a resurgent green technology to do with the problem of rural poverty? In the past, green technology has always been rural, based in farms and villages rather than in cities. In the future it will pervade cities as well as the countryside, factories as well as forests. It will not be entirely rural. But it will still have a large rural component. After all, the cloning of Dolly occurred in a rural animal-breeding station in Scotland, not in an urban laboratory in Silicon Valley. Green technology will use land and sunlight as its primary sources of raw materials and energy. Land and sunlight cannot be concentrated in cities but are spread more or less evenly over the planet. When industries and technologies are based on land and sunlight, they will bring employment and wealth to rural populations.

In a country like India with a large rural population, bringing wealth to the villages means bringing jobs other than farming. Most of the villagers must cease to be subsistence farmers and become shopkeepers or schoolteachers or bankers or engineers or poets. In the end the villages must become gentrified, as they are today in England, with the old farm workers' cottages converted into garages, and the few remaining farmers converted into highly skilled professionals. It is fortunate that sunlight is most abundant in tropical countries, where a large fraction of the world's people live and where rural poverty is most acute. Since sunlight is distributed more equitably than coal and oil, green technology can be a great equalizer, helping to narrow the gap between rich and poor countries.

My book *The Sun, the Genome, and the Internet* (1999) describes a vision of green technology enriching villages all over the world and halting the migration from villages to megacities. The three components of the vision are all essential: the sun to provide energy where it is needed, the genome to provide plants that can convert sunlight into chemical fuels cheaply and efficiently, the Internet to end the intellectual and economic isolation of rural populations. With all three components in place, every village in Africa could enjoy its fair share of the blessings of civilization. People who prefer to live in cities would still be free to move from villages to cities, but they would not be compelled to move by economic necessity.

STEVE FEATHERSTONE

The Coming Robot Army

FROM *Harper's Magazine*

A SMALL GRAY HELICOPTER was perched on the runway, its rotors beating slowly against the shroud of fog and rain blowing in from Chesapeake Bay. Visibility was poor, but visibility did not matter. The helicopter had no windows, no doors, and, for that matter, no pilot. Its elliptical fuselage looked as if it had been carved out of wood and sanded smooth of detail. It hovered above the runway for a moment, swung its blind face toward the bay, and then dissolved into the mist.

The helicopter was the first among a dozen unmanned aerial vehicles (UAVs) scheduled to fly during the annual Association for Unmanned Vehicle Systems International conference in Baltimore. The live demonstration area at Webster Field, a naval air facility located seventy miles south of Washington, D.C., was laid out along the lines of a carnival midway. Big defense contractors and small engineering firms exhibited the latest military robots under white tents staked out alongside an auxiliary runway. Armed soldiers kept watch from towers and strolled through the throng of military officers and industry reps. I took a seat among rows of metal chairs arrayed in front of a giant video screen, which displayed a live feed from the helicopter's surveillance camera. There was little to see except clouds, so the announcer attempted to liven things up.

"Yesterday we saw some boats out there," he said, with an aggressive enthusiasm better suited to a monster-truck rally. "They didn't know they were being targeted by one of the newest UAVs!"

Next, two technicians from AeroVironment, Inc., jogged onto

the airfield and knelt in the wet grass to assemble what appeared to be a remote-controlled airplane. One of them raised it over his shoulder, leaned back, and threw it into the air like a javelin. The airplane, called the Raven, climbed straight up, stalled, dipped alarmingly toward the ground, and then leveled off at two hundred feet, its tiny electric motor buzzing like a mosquito. The screen switched to show the Raven's video feed: a bird's-eye view of the airstrip, at one end of which a large American flag flapped limply on a rope strung between two portable cranes next to an inflatable Scud missile launcher.

"A lot of the principles we use here are taken from the model industry," an AeroVironment spokesman told the announcer as the Raven looped around the field. The U.S. military has purchased more than 3,000 Ravens, many of which have been deployed in Iraq and Afghanistan, but apparently none of the military officers present had ever seen one land. At the end of the Raven's second flight, the crowd went silent as the tiny plane plummeted from the sky and careered into the ground, tearing off its wings. The technicians scrambled to the crash site, stuck the wings back on, and held the Raven triumphantly above their heads.

"It's designed that way," the spokesman explained.

"Hey, if you can't fix it with duct tape," the announcer said, "it's not worth fixing, am I right?"

Other teams took the field to demonstrate their company's UAVs. The sheer variety of aircraft and their launching methods — planes were slung from catapults and bungee cords, shot from pneumatic guns and the backs of pickup trucks, or simply tossed by hand into the air — testified to the prodigious growth in demand for military robots since the terrorist attacks of September 11, 2001, and the subsequent "global war on terrorism." In his opening conference remarks, Rear Admiral Timothy Heely compared the embryonic UAV market with aviation in the first decades of the twentieth century, when the Wright brothers built planes in their workshop and dirigibles carried passengers. "It's all out there," he said. "You don't want to throw anything away."

It started to drizzle again. The military officers sought refuge under a catered VIP tent decorated with red, white, and blue bunting while the rest of us scattered in all directions. I headed to the unmanned ground vehicle (UGV) tent, located at the far end of the runway. The tent's interior was dim; the air, sticky and hot. Tables stocked with brochures and laptops lined the vinyl walls. Robots rested unevenly on the grass. This was the first year UGVs were allowed demonstration time at the conference, and company reps were eager to show what their robots could do. A rep from iRobot, maker of the popular Roomba robotic vacuum cleaner, flipped open a shiny metal briefcase that contained an LCD monitor and a control panel studded with switches and buttons for operating the PackBot, a "man-packable" tracked robot not much bigger than a telephone book. Hundreds of PackBots have already been deployed in Iraq.

"If you can operate a Game Boy, you're good," the rep said.

A Raytheon engineer fired up an orange robot that looked like a track loader used in excavation. The only difference was a solid black box containing a radio receiver on top of the cage where the human driver normally sat. It rumbled out of the tent onto the airfield, followed by a camera crew.

"It's a Bobcat," the announcer shouted. "It's a *biiig* Bobcat!"

The Bobcat rolled up to a steel garbage bin containing a "simulated Improvised Explosive Device," hoisted it into the air with a set of pincers, and crumpled it like a soda can. A Raytheon spokesman listed all the things the tricked-out Bobcat could do, such as breach walls.

"You could also crush things like a car if you wanted to," he added.

"I never thought of crushing something," the announcer said. "But yeah, this would do very nicely."

After the Bobcat had dispatched the mangled garbage bin and returned to the tent, I asked a Raytheon engineer if the company had thought about arming it with machine guns. "Forget the machine guns," he said dismissively. "We're going lasers."

Military robots are nothing new. During World War II, Germans sent small, remote-controlled bombs on tank treads across front lines, and the United States experimented with unmanned aircraft, packing tons of high explosives into conventional bombers piloted from the air by radio (one bomber exploded soon after takeoff, killing Joseph Kennedy's eldest son, and the experiment was eventually shelved). But in a war decided by the maneuver of vast armies across whole continents, robots were a peculiar sideshow.

The practice of warfare has changed dramatically in the past sixty years. Since Vietnam, the American military machine has been governed by two parallel and complementary trends: an aversion to casualties and a heavy reliance on technology. The Gulf War reinforced the belief that technology can replace human soldiers on the battlefield, and the "Black Hawk down" incident in Somalia made this belief an article of faith. Today, any new weapon worth its procurement contract is customarily referred to as a "force multiplier," which can be translated as doing more damage with less people. Weaponized robots are the ultimate force multiplier, and every branch of the military has increased spending on new unmanned systems.

At $145 billion, the army's Future Combat Systems (FCS) is the costliest weapons program in history, and in some ways the most visionary as well. The individual soldier is still central to the FCS concept, but he has been reconfigured as a sort of plug-and-play warrior, a node in what is envisioned as a sprawling network of robots, manned vehicles, ground sensors, satellites, and command centers. In theory, each node will exchange real-time information with the network, allowing the entire system to accommodate sudden changes in the "battle space." The fog of war would become a relic of the past, like the musket, swept away by crystalline streams of encrypted data. The enemy would not be killed so much as deleted.

FCS calls for seven new unmanned systems. It's not clear how much autonomy each system will be allowed. According to *Unmanned Effects (UFX): Taking the Human out of the Loop,* a 2003 study commissioned by the U.S. Joint Forces Command, advances in artificial intelligence and automatic target recognition will give robots the ability to hunt down and kill the enemy with limited human supervision by 2015. As the study's title suggests, humans are the weakest link in the robot's "kill chain" — the sequence of events that occurs from the moment an enemy target is detected to its destruction.

At Webster Field, the latest link in the military's increasingly automated kill chain was on display: the Special Weapons Observation Reconnaissance Detection System, or SWORDS. I squatted down to take a closer look at it. Despite its theatrical name, SWORDS was remarkably plain, consisting of two thick rubber treads, stubby antennae, and a platform mounted with a camera and an M240 machine gun, all painted black. The robot is manu-

factured by a company named Foster-Miller, whose chief representative at the show was Bob Quinn, a slope-shouldered, balding man
with bright blue eyes. Bob helped his engineer get SWORDS ready
for a quick demo. Secretary of the Army Francis Harvey, the VIP of
VIPs, was coming through the UGV tent for a tour.

"The real demonstration is when you're actually firing these
things," Bob lamented. Unfortunately, live fire was forbidden at
Webster Field, and Bob had arrived too late to schedule a formal
demonstration. At another conference two months before, he had
been free to drive SWORDS around all day long. "I was going into
the different booths and displays, pointing my gun, moving it up
and down like the sign of the cross. People were going like this" —
he jumped back and held up his hands in surrender — "then they
would follow the robot back to me because they had no idea where
I was. And that's the exact purpose of an urban combat capability
like this."

Sunlight flooded into the tent as Secretary Harvey parted the
canopy, flanked by two lanky Rangers in fatigues and berets. Bob
ran his hand over his scalp and smoothed his shirt. It was sweltering
inside the tent now. Beneath the brim of his tan baseball cap, Secretary Harvey's face was bright red and beaded with sweat. He nodded politely, leaning into the verbal barrage of specifications and
payloads and mission packages the reps threw at him. When he got
to SWORDS, he clasped his hands behind his back and stared
down at the robot as if it were a small child. Someone from his entourage excitedly explained the various weapons it could carry.

Bob had orchestrated enough dog-and-pony shows to know that
technology doesn't always impress men of Secretary Harvey's age
and position. "We don't have it in the field yet," Bob interrupted,
going on to say that SWORDS wasn't part of any official procurement plan. It was a direct result of a "bootstrap effort" by real soldiers at Picatinny Arsenal in New Jersey who were trying to solve
real problems for their comrades in the field. "And soldiers love it,"
he added.

On the long bus ride back to Baltimore, I sat behind Master Sergeant Mike Gomez, a Marine UAV pilot. "All we are are battery-
powered forward observers," he joked. Mike was biased against autonomous robots that could fire weapons or drop bombs with minimal, if any, human intervention. There were too many things that

could go wrong, and innocent people could be killed as a result. At the same time, he wasn't opposed to machines that were "going to save Marines, save time, save manpower, save lives."

It wasn't the first time that day I'd heard this odd contradiction, and over the next three days I'd hear it again and again. It was as if everyone had rehearsed the same set of talking points. Robots will take soldiers out of harm's way. Robots will save lives. Allow robots to pull the trigger? No way, it'll never happen. But wasn't the logical outcome of all this fancy technology an autonomous robot force, no humans required save for those few sitting in darkened control rooms half a world away? Wasn't the best way to save lives — American lives, at least — to take humans off the battlefield altogether? Mike stared out the bus window at the passing traffic.

"I don't think that you can ever take him out," he said, his breath fogging the tinted glass. "What happens to every major civilization? At some point they civilize themselves right out of warriors. You've got sheep and you've got wolves. You've got to have enough wolves around to protect your sheep, or else somebody else's wolves are going to take them out."

Coming from a career soldier, Mike's views of war and humanity were understandably romantic. To him, bad wolves weren't the real threat. It was the idea that civilization might be able to get along without wolves, good or bad, or that wolves could be made of titanium and silicon. What would happen to the warrior spirit then?

Scores of scale-model UAVs dangled on wires from the ceiling of the exhibit hall at the Baltimore Convention Center, rotating lazily in currents of air conditioning. Models jutted into the aisles, their wings canted in attitudes of flight. Company reps blew packing dust off cluster bombs and electronic equipment. They put out bowls of candy and trinkets. Everywhere I looked I saw ghostly black-and-white images of myself, captured by dozens of infrared surveillance cameras mounted inside domed gimbals, staring back at me from closed-circuit televisions.

In addition to cameras, almost every booth featured a large plasma monitor showing a continuous video loop of robots blowing up vehicles on target ranges, robots pepper-spraying intruders, robots climbing stairs, scurrying down sewer pipes, circling above battlefields and mountain ranges. These videos were often accompanied by a narrator's bland voice-over, muttered from a sound system that rivaled the most expensive home theater.

I sat down in the concession area to study the floor map. An engineer next to me picked at a plate of underripe melon and shook his head in awe at the long lines of people waiting for coffee. "Four or five years ago it was just booths with concept posters pinned up," he said. "Now the actual stuff is here. It's amazing."

At the fringes of the exhibit hall, I wandered through the warrens of small companies and remote military arsenals squeezed side by side into ten-by-ten booths. I followed the screeching chords of thrash metal until I stood in front of a television playing a promotional video featuring a robot called Chaos. Chaos was built by Autonomous Solutions, a private company that had been spun out of Utah State University's robotics lab. In the video, it clambered over various types of terrain, its four flipperlike tracks chewing up dirt and rocks and tree bark. The real thing was somewhat less kinetic. A Chaos prototype lay motionless on the floor in front of the television. I nudged it with my foot and asked the company's young operations manager what it was designed to do.

"Kick the pants off the PackBot," he said, glancing around nervously. "No, I'm kidding."

A few booths down I encountered a group of men gathered around a robot the size of a paperback book. Apparently, it could climb walls by virtue of a powerful centrifuge in its belly. A picture showed it stuck to a building outside a second-story window, peering over the sill. But the rep holding the remote-control box kept ramming the robot into a cloth-draped wall at the back of his booth. The robot lost traction on the loose fabric and flipped over on its back, wheels spinning. A rep from the neighboring booth volunteered the use of his filing cabinet. The little robot zipped across the floor, bumped the cabinet, and, with a soft whir, climbed straight up the side. When it got to the top it extended a metal stalk bearing a tiny camera and scanned the applauding crowd.

I continued along the perimeter, trying to avoid eye contact with the reps. Since it was the first day of the show, they were fresh and alert, rocking on their heels at the edges of their booths, their eyes darting from name badge to name badge in search of potential customers. I picked up an M4 carbine resting on a table in the Chatten Associates booth. The gun's grip had been modified to simulate a computer mouse. It had two rubber keys and a thumb stick for operating a miniature radio-controlled tank sporting an assault rifle in its turret.

"You'll need this," said Kent Massey, Chatten's chief operating officer. He removed a helmet from a mannequin's head and placed it on mine. Then he adjusted the heads-up display, a postage stamp–size LCD screen that floated in front of my right eye. The idea behind the setup was that a soldier could simultaneously keep one eye on the battlefield while piloting the robot via a video feed beamed back to his heads-up display. He never had to take his finger off the trigger.

I blinked and saw a robot's-eye view of traffic cones arranged on a fluorescent green square of artificial turf. I turned my head first to the left, then to the right. The gimbal-mounted camera in the tank mimicked the motion, swiveling left, then right. I pushed the thumb stick on the carbine's pistol grip. The tank lurched forward, knocking down a cone.

"Try not to look at the robot," Kent advised.

I turned my back to him and faced the aisle. It was difficult for me to imagine how the soldier of the future would manage both the stress of combat and the information overload that plagues the average office worker. Simply driving the tank made me dizzy, despite Kent's claims that Chatten's head-aiming system increased "situational awareness" and "operational efficiency" by 400 percent. Then again, I wasn't army material. I was too old, too analog. As a Boeing rep would later explain to me, they were "building systems for kids that are in the seventh and eighth grades right now. They get the PDAs, the digital things, cell phones, IM."

As I crashed the tank around the obstacle course, conventioneers stopped in the aisle to determine why I was pointing a machine gun at them. I aimed the muzzle at the floor.

"The one mission that you simply cannot do without us is armed reconnaissance," Kent said over my shoulder. "Poke around a corner, clear a house . . . We lost thirty-eight guys in Fallujah in exactly those kinds of circumstances, plus a couple hundred wounded. If [the robot] gets killed, there's no letter to write home."

Robots have always been associated with dehumanization and, more explicitly, humanity's extinction. The word "robot" is derived from the Czech word for forced labor, *vobota*, and first appeared in Karel Capek's 1920 play, *R.U.R. (Rossum's Universal Robots)*, which ends with the destruction of mankind.

This view of robots, popularized in such movies as the Terminator series, troubles Cliff Hudson, who at the time coordinated robotics efforts for the Department of Defense. I ran into Cliff on the second day of the show, outside Carnegie Mellon's National Robotics Engineering Center's booth. Like the scientists in *R.U.R.*, Cliff saw robots as a benign class of mechanized serfs. Military robots will handle most of "the three Ds: dull, dangerous, dirty-type tasks," he said, such as transporting supplies, guarding checkpoints, and sniffing for bombs. The more delicate task of killing would remain in human hands.

"I liken it to the military dog," Cliff said, and brought up a briefing given the previous day by an explosive-ordnance disposal (EOD) officer who had just returned from Iraq. The highlight of the briefing was an MTV-style video montage of robots disarming IEDs. It ended with a soldier walking away from the camera, silhouetted against golden evening sunlight, his loyal robot bumping along the road at his heels. Cliff pressed his hands together. "It's that partnership, it's that team approach," he said. "It's not going to replace the soldier. It's going to be an added capability and enhancer."

Adjacent to where we stood talking in the aisle was a prototype of the Gladiator, a six-wheeled armored car about the size of a golf cart, built by Carnegie Mellon engineers for the Marines. It was one mean enhancer. The prototype was equipped with a machine gun, but missiles could be attached to it as well.

"If you see concertina wire, you send this downrange," Cliff said, maintaining his theme of man/robot cooperation. "And then the Marines can come up behind it. It's a great weapon." Despite its capabilities, the Gladiator hadn't won the complete trust of the Marines. "It's a little unstable," Cliff admitted. "Most people are uncomfortable around it when the safety is removed."

Reps proffering business cards began circling around Cliff and his entourage, sweeping me aside. Jörgen Pedersen, a young engineer with thin blond hair and a goatee, watched the scene with bemused detachment, his elbows propped on the Gladiator's turret. Jörgen had written the Gladiator's fire-control software.

"How safe is this thing?" I asked him.

"We wanted it to err on the side of safety first," Jörgen said. "You can always make something more unsafe." In the early stages of the Gladiator's development, Jörgen had discovered that its communications link wasn't reliable enough to allow machine-gun bursts longer than six seconds. After six seconds, the robot would stop firing. So he reprogrammed the fire-control system with a fail-safe.

"You may have great communications here," Jörgen said, touching the Gladiator with his fingertips. "But you take one step back and you're just on the hairy edge of where this thing can communicate well."

The integrity of data links between unmanned systems and their operators is a major concern. Satellite bandwidth, already in short supply, will be stretched even further as more robots and other sophisticated electronics, such as remote sensors, are committed to the battlefield. There's also the possibility that radio signals could be jammed or hijacked by the enemy. But these problems are inherent to the current generation of teleoperated machines: robots that are controlled by humans from afar. As robots become more autonomous, fulfilling missions according to preprogrammed instructions, maintaining constant contact with human operators will be unnecessary. I asked Jörgen if robots would someday replace soldiers on the battlefield. He reiterated the need for a man in the loop.

"Maybe that's because I'm shortsighted based on my current experiences," he said. "Maybe the only way that it could happen is if there're no other people out on that field doing battle. It's just robots battling robots. At that point, it doesn't matter. We all just turn on the TV to see who's winning."

It is almost certain that robot deployment will save lives, both military and civilian. And yet the prospect of robot-on-human warfare does present serious moral and ethical, if not strictly legal, issues. Robots invite no special consideration under the laws of armed conflict, which place the burden of responsibility on humans, not weapons systems. When a laser-guided bomb kills civilians, responsibility falls on everyone involved in the kill chain, from the pilot who dropped the bomb to the commander who ordered the strike. Robots will be treated no differently. It will become vastly more difficult, however, to assign responsibility for noncombatant deaths caused by mechanical or programming failures as robots are granted greater degrees of autonomy. In this sense, robots may prove similar to low-tech cluster bombs or land mines, munitions that "do something that they're not supposed to out of the control of those who deploy them, and in doing so cause unintended death and suffering," according to Michael Byers, professor of global politics and international law at the University of British Columbia.

The moral issues are perhaps similar to those arising from the use of precision-guided munitions (PGMs). There's no doubt that PGMs greatly limit civilian casualties and collateral damage to civilian infrastructure such as hospitals, electrical grids, and water systems. But because PGM strikes are more precise compared with dropping sticks of iron bombs from B-52s, the civilian casualties that often result from PGM strikes are considered necessary, if horribly unfortunate, mistakes. One need look no further than the PGM barrage that accompanied the ground invasion of Iraq in 2003. "Decapitation strikes" aimed at senior Iraqi leaders pounded neighborhoods from Baghdad to Basra. Due to poor intelligence, none of the fifty known strikes succeeded in finding their targets. In four of the strikes forty-two civilians were killed, including six members of a family who had the misfortune of living next door to Saddam Hussein's half brother.

It's not difficult to imagine a similar scenario involving robots instead of PGMs. A robot armed only with a machine gun enters a house known to harbor an insurgent leader. The robot opens fire and kills a woman and her two children instead. It's later discovered that the insurgent leader moved to a different location at the last minute. Put aside any mitigating factors that might prevent a situation like this from occurring and assume that the robot did exactly what it was programmed to do. Assume the commander behind the operation acted on the latest intelligence and that he followed the laws of armed conflict to the letter. Although the deaths of the woman and children might not violate the laws of armed conflict, they fall into a moral black hole where no one, no human anyway, is directly responsible. Had the innocents of My Lai and Haditha been slain not by errant men but by errant machines, would we know the names of these places today?

More troubling than the compromised moral calculus with which we program our killing machines is how robots reduce even further the costs, both fiscal and human, of the choice to wage war. Robots do not have to be recruited, trained, fed, or paid extra for combat duty. When they are destroyed, there are no death benefits to disburse. Shipping them off to hostile lands doesn't require the expenditure of political capital either. There will be no grieving robot mothers pitching camp outside the president's ranch gates. Robots are, quite literally, an off-the-shelf war-fighting capability — war in a can.

This bloodless vision of future combat was best captured by a billboard I saw at the exhibition, in the General Dynamics booth. The billboard was titled ROBOTS AS CO-COMBATANTS, and two scenes illustrated the concept in the garish style of toy-model-box art. One featured UGVs positioned on a slope near a grove of glossy palm trees. In the distance, a group of mud-brick buildings resembling a walled compound was set against a barren mountain range. Bright red parabolas traced the trajectories of mortar shells fired into the compound from UGVs, but there were no explosions, no smoke.

The other scene was composed in the gritty vernacular of television news footage from Iraq. A squad of soldiers trotted down the cracked sidewalk of a city street, past stained concrete façades and terraces awash in glaring sunlight. A small, wingless micro-UAV hovered above the soldiers amid a tangled nest of drooping telephone lines, projecting a cone of white light that suggested an invisible sensor beam. And smack in the foreground, a UGV had maneuvered into the street, guns blazing. In both scenes, the soldiers were incidental to the action. Some didn't even carry rifles. They sat in front of computer screens, fingers tapping on keyboards.

On the last day of the show, I sat in the concession area, chewing a stale pastry and scanning the list of the day's technical sessions. Most were dry, tedious affairs with such titles as "The Emerging Challenge of Loitering Attack Missiles." One session hosted by Foster-Miller, the company that manufactures the SWORDS robot, got my attention: "Weaponization of Small Unmanned Ground Vehicles." I filled my coffee cup and hustled upstairs.

I took a seat near the front of the conference room just as the lights dimmed. Hunched behind a podium, a Foster-Miller engineer began reading verbatim from a PowerPoint presentation about the history of SWORDS, ending with a dreary bullet-point list cataloguing the past achievements of the TALON robot, SWORDS's immediate predecessor.

"TALON has been used in most major, major . . ." The engineer faltered.

"Conflicts," someone in the audience stage-whispered. I turned to see that it was Bob Quinn. He winked at me in acknowledgment.

"Conflicts," the engineer said. He ended his portion of the talk with the same video montage that had inspired Cliff Hudson to compare robots to dogs. TALON robots were shown pulling apart tangles of wire connected to IEDs, plucking at garbage bags that had been tossed on the sides of darkened roads, extracting mortar shells hidden inside Styrofoam cups. Bob Quinn took the podium just as the final shot in the montage, that of the soldier walking down the road with his faithful TALON robot at his heels, faded on the screen behind him. The lights came up.

"The eight-hundred-pound gorilla, or the bully in the playpen, for weaponized robotics — for all ground-based robots — is Hollywood," Bob said. The audience stirred. Bob strolled off the dais and stood in the aisle, hands in his pockets. "It's interesting that UAVs like the Predator can fire Hellfire missiles at will without a huge interest worldwide. But when you get into weaponization of ground vehicles, our soldiers, our safety community, our nation, our world, are not ready for autonomy. In fact, it's quite the opposite."

Bob remained in the aisle, narrating a series of PowerPoint slides and video clips that showed SWORDS firing rockets and machine guns, SWORDS riding atop a Stryker vehicle, SWORDS creeping up on a target and lobbing grenades at it. His point was simple: SWORDS was no killer robot, no Terminator. It was a capable weapons platform firmly in the control of the soldiers who operated it, nothing more. When the last video clip didn't load, Bob stalled for time.

"We've found that using Hollywood on Hollywood is a good strategy to overcome some of the concerns that aren't apparent with UAVs but are very apparent with UGVs," he said. Last February a crew from the History Channel had filmed SWORDS for an episode of *Mail Call*, a half-hour program hosted by the inimitable

R. Lee Ermey, best known for his role as the profane drill sergeant in the movie *Full Metal Jacket.* Ermey's scowling face suddenly appeared onscreen, accompanied by jarring rock music.

"It's a lot smarter to send this robo-soldier down a blind alley than one of our flesh-and-blood warriors," Ermey shouted. "It was developed by our troops in the field, not some suit in an office back home!"

Ermey's antic mugging was interspersed with quick cutaways of SWORDS on a firing range and interviews with EOD soldiers.

"The next time you start thinking about telling the kids to put away that video game, think again!" Ermey screamed. He jabbed his finger into the camera. "Someday they could be using those same kinds of skills to run a robot that will save their bacon!"

"That's a good way to get off the stage," Bob said. He was smiling now, soaking in the applause. "I think armed robots will save soldiers' lives. It creates an unfair fight, and that's what we want. But they will be teleoperated. The more as a community we focus on that, given the Hollywood perceptions, the better off our soldiers will be."

Downstairs in the exhibit hall, I saw that Boeing had also learned the value of Hollywood-style marketing. I had stopped by the company's booth out of a sense of obligation more than curiosity: Boeing is the lead contractor for FCS. While I was talking to Stephen Bishop, the FCS business-development manager, I noticed a familiar face appear on the laptop screen behind him.

"Is that — MacGyver?"

Stephen nodded and stepped aside so that I could get a better view of the laptop. The face did indeed belong to Richard Dean Anderson, former star of the television series *MacGyver* and now the star of a five-minute promotional film produced by Boeing. Judging by the digital special effects, the film had probably cost more to make than what most companies had spent on their entire exhibits. Not coincidentally, the film is set in 2014, when the first generation of FCS vehicles is scheduled for full deployment. An American convoy approaches a bridge near a snowy mountain pass somewhere in Asia, perhaps North Korea. The enemy mobilizes to cut the Americans off, but they are detected and annihilated by armed ground vehicles and UAVs.

At the center of this networked firestorm is Richard Dean Anderson, who sits inside a command vehicle, furrowing his brow and tapping a computer touchscreen. As the American forces cross the bridge, a lone enemy soldier hiding behind a boulder fires a rocket at the lead vehicle and disables it. The attack falters.

"I do *not* have an ID on the shooter!" a technician yells. Anderson squints grimly at his computer screen. It's the moment of truth. Does he pull back and allow the enemy time to regroup or does he advance across the bridge, exposing his forces to enemy fire? The rousing martial soundtrack goes quiet.

"Put a 'bot on the bridge," Anderson says.

A dune buggy–like robot darts from the column of vehicles and stops in the middle of the bridge in a heroic act of self-sacrifice. The lone enemy soldier takes the bait and fires another missile, destroying the robot and unwittingly revealing his position to a micro-UAV loitering nearby. Billions of dollars and decades of scientific research come to bear on this moment, on one man hiding behind a snow-covered boulder. He is obliterated.

"Good job," Anderson sneers. "Now let's finish this."

The film ends as American tanks pour across the bridge into enemy territory. The digitally enhanced point of view pulls back to reveal the FCS network, layer by layer, vehicle by vehicle, eighteen systems in all, until it reaches space, the network's outer shell, where a spy satellite glides by.

"Saving soldiers' lives," Stephen said, glancing at his press manager to make sure he was on message. I commended the film's production values. Stephen seemed pleased that I'd noticed. "Three-stars and four-stars gave it a standing ovation at the Pentagon last November," he told me.

"You can't argue with MacGyver," I said.

"Because it's all about saving soldiers' lives," Stephen said. "Works for congressmen, works for senators, works for the grandmother in Nebraska."

Later that summer I visited Picatinny Arsenal, "Home of American Firepower," in New Jersey, to see a live-fire demonstration of the SWORDS robot. SWORDS was conceived at Picatinny by a small group of EOD soldiers who wanted to find a less dangerous way to "put heat on a target" inside caves in Afghanistan. Three years later, SWORDS was undergoing some final tweaks at Picatinny be-

fore being sent to Aberdeen Proving Ground for its last round of
safety tests. After that, it would be ready for deployment.

"As long as you don't break my rules you'll be fine," said Sergeant Jason Mero, motioning for us to gather around him. Sergeant Mero had participated in the initial invasion of Iraq, including the assault on Saddam International Airport. He had buzzed
sandy brown hair, a compact build, and the brusque authority common to noncommissioned officers. He told us exactly where we
could stand, where we could set up our cameras, and assured us
that he was there to help us get what we needed. Other than the
"very, very loud" report of the M240 machine gun, there was little
to worry about.

"The robot's not going to suddenly pivot and start shooting everybody," he said, without a hint of irony.

A crew from the Discovery Networks' Military Channel dragged
their gear onto the range. They were filming a special on warbots,
and the producer was disappointed to learn that the SWORDS robot mounted with a formidable-looking M202 grenade launcher
wasn't operable. He would have to make do with the less telegenic
machine-gun variant. The producer, Jonathan Gruber, wore a canvas fishing hat with the brim pulled down to the black frames of his
stylish eyeglasses. Jonathan gave stage directions to Sergeant Mero,
who knelt in the gravel next to SWORDS and began describing
how the loading process works.

"Sergeant, if you could just look to me," Jonathan prompted.
"Good. So, is a misfeed common?"

"No, not with this weapon system," Sergeant Mero said. "It's very
uncommon."

"My questions are cut out," Jonathan said. "So if you could repeat
my question in the answer? So, you know, 'Misfeeds are not common . . .'"

"Mis —" Sergeant Mero cleared his throat. His face turned red.
"However, misfeeds are not common with the M240 Bravo."

"Okay, great. I'm all set for now, thanks."

The firing range was scraped out of the bottom of a shallow
gorge, surrounded on all sides by trees and exposed limestone.
Turkey vultures circled above the ridge. The weedy ground was littered with spent shell casings and scraps of scorched metal. Fifty
yards from where I sat, two human silhouettes were visible through

shoulder-high weeds in front of a concrete trap filled with sand. Sergeant Mero hooked a cable to SWORDS's camera, then flipped a red switch on the control box. I felt the M240's muzzle blast on my face as SWORDS lurched backward on its tracks, spilling smoking shells on the ground.

A cloud of dust billowed behind the silhouettes. Sergeant Mero fired again, then again. With each burst, recoil pushed SWORDS backward, and Sergeant Mero, staring at the video image on the control box's LCD screen, readjusted his aim. I could hear servos whining. When Sergeant Mero finished the ammunition belt, he switched off SWORDS and led us downrange to the targets.

"So, um, Sergeant?" Jonathan said. "As soon as you see our camera you can just start talking."

"As you see, the M240 —"

"And Sergeant?" Jonathan interrupted. "I don't think you have to scream. You can just speak in a normal voice. We're all close to you."

"The problem with a heavy machine gun is, obviously, there's going to be a lot of spray," Sergeant Mero said, bending down to pick up one of the silhouettes that had fallen in the weeds. "Our second guy over here that we actually knocked down — he didn't get very many bullets, but he actually got hit pretty hard."

Through the weeds I spotted the SWORDS robot squatting in the dust. My heart skipped a beat. The machine gun was pointed straight at me. I'd watched Sergeant Mero deactivate SWORDS. I saw him disconnect the cables. And the machine gun's feed tray was empty. There wasn't the slightest chance of a misfire. My fear was irrational, but I still made a wide circle around the robot when it was time to leave.

Within our lifetime, robots will give us the ability to wage war without committing ourselves to the human cost of actually fighting a war. War will become a routine, a program. The great nineteenth-century military theorist Carl von Clausewitz understood that although war may have rational goals, the conduct of war is fundamentally irrational and unpredictable. Absent fear, war cannot be called war. A better name for it would be target practice.

Back on the firing line, Sergeant Mero booted up SWORDS and began running it around the range for the benefit of the cameras.

It made a tinny, rattling noise as it rumbled over the rocks. A Discovery crewman waddled close behind it, holding his camera low to the ground. He stumbled over a clump of weeds, and for a second I thought he was going to fall on his face. But he regained his balance, took a breath, and ran to catch up with the robot.

"I think I'm good," Jonathan said after the driving demonstration. "Anything else you want to add about this?"

"Yeah," Sergeant Mero said, smiling wryly. "It kicks *ass*. It's *awesome*." In repentance for this brief moment of sarcasm, Sergeant Mero squared his shoulders, looked straight into the camera, and began speaking as if he were reading from cue cards. "These things are amazing," he said breathlessly. "They don't complain, like our regular soldiers do. They don't cry. They're not scared. This robot here has no fear, which is a good supplement to the United States Army."

"That's great," Jonathan said.

MICHAEL FINKEL

Malaria: Stopping a Global Killer

FROM *National Geographic*

IT BEGINS WITH A BITE, a painless bite. The mosquito comes in the night, alights on an exposed patch of flesh, and assumes the hunched, head-lowered posture of a sprinter in the starting blocks. Then she plunges her stiletto mouthparts into the skin.

The mosquito has long, filament-thin legs and dappled wings; she's of the genus *Anopheles,* the only insect capable of harboring the human malaria parasite. And she's definitely a she: male mosquitoes have no interest in blood, while females depend on protein-rich hemoglobin to nourish their eggs. A mosquito's proboscis appears spike-solid, but it's actually a sheath of separate tools — cutting blades and a feeding tube powered by two tiny pumps. She drills through the epidermis, then through a thin layer of fat, then into the network of blood-filled microcapillaries. She starts to drink.

To inhibit the blood from coagulating, the mosquito oils the bite area with a spray of saliva. This is when it happens. Carried in the mosquito's salivary glands — and entering the body with the lubricating squirt — are minute, wormlike creatures. These are the one-celled malaria parasites known as plasmodia. Fifty thousand of them could swim in a pool the size of the period at the end of this sentence. Typically, a couple of dozen slip into the bloodstream. But it takes just one. A single plasmodium is enough to kill a person.

The parasites remain in the bloodstream for only a few minutes.

They ride the flume of the circulatory system to the liver. There they stop. Each plasmodium burrows into a different liver cell. Almost certainly, the person who has been bitten hardly stirs from sleep. And for the next week or two, there's no overt sign that something in the body has just gone horribly wrong.

We live on a malarious planet. It may not seem that way from the vantage point of a wealthy country, where malaria is sometimes thought of, if it is thought of at all, as a problem that has mostly been solved, like smallpox or polio. In truth, malaria now affects more people than ever before. It's endemic to 106 nations, threatening half the world's population. In recent years, the parasite has become so entrenched and has developed resistance to so many drugs that the most potent strains can scarcely be controlled. This year malaria will strike up to half a billion people. At least a million will die, most of them under the age of five, the vast majority living in Africa. That's more than twice the annual toll a generation ago.

The outcry over this epidemic, until recently, has been muted. Malaria is a plague of the poor, easy to overlook. The most unfortunate fact about malaria, some researchers believe, is that prosperous nations got rid of it. In the meantime, several distinctly unprosperous regions have reached the brink of total malarial collapse, virtually ruled by swarms of buzzing, flying syringes.

Only in the past few years has malaria captured the full attention of aid agencies and donors. The World Health Organization (WHO) has made malaria reduction a chief priority. Bill Gates, who has called malaria "the worst thing on the planet," has donated hundreds of millions of dollars to the effort through the Bill and Melinda Gates Foundation. The Bush administration has pledged $1.2 billion. Funds devoted to malaria have doubled since 2003. The idea is to disable the disease by combining virtually every known malaria-fighting technique, from the ancient (Chinese herbal medicines) to the old (bed nets) to the ultramodern (multidrug cocktails). At the same time, malaria researchers are pursuing a long-sought, elusive goal: a vaccine that would curb the disease for good.

Much of the aid is going to a few hard-hit countries scattered across sub-Saharan Africa. If these nations can beat back the disease, they'll serve as templates for the global antimalaria effort.

And if they can't? Well, nobody in the malaria world likes to answer that question.

One of these spotlighted countries — perhaps the place most closely watched by malaria experts — is Zambia, a sprawling, land-locked nation carved out of the fertile bushland of southern Africa. It's difficult to comprehend how thoroughly Zambia has been devastated by malaria. In some provinces, at any given moment more than a third of all children under age five are sick with the disease.

Worse than the sheer numbers is the type of malaria found in Zambia. Four species of malaria parasites routinely infect humans; the most virulent, by far, is *Plasmodium falciparum*. About half of all malaria cases worldwide are caused by *falciparum,* and 95 percent of the deaths. It's the only form of malaria that can attack the brain. And it can do so with extreme speed — few infectious agents can overwhelm the body as swiftly as *falciparum*. An African youth can be happily playing soccer in the morning and be dead of *falciparum* malaria that night.

Falciparum is a major reason that nearly 20 percent of all Zambian babies do not live to see their fifth birthday. Older children and adults, too, catch the disease — pregnant women are especially prone — but most have developed just enough immunity to fight the parasites to a stalemate, though untreated malaria can persist for years, the fevers fading in and out. There are times when it seems that everyone in Zambia is debilitated to some degree by malaria; many have had it a dozen or more times. No surprise that the nation remains one of the poorest in the world: a country's economic health has little chance of improving until its physical health is revitalized. Zambia's goal is to reduce malaria deaths by 75 percent over the next four years.

To witness the full force of malaria's stranglehold on Zambia, it's essential to leave the capital city, Lusaka. Drive north, across the verdant plains, past the banana plantations and the copper mines — copper is Zambia's primary export — and into the forested region tucked between the borders of Angola and the Democratic Republic of the Congo. This is North-Western Province. It is almost entirely rural; many villages can be reached only by thin footpaths worn into the beet-red soil. A nationwide health survey in 2005 concluded that for every 1,000 children under age five living in North-Western Province, there were 1,353 cases of malaria. An annual rate of more than 100 percent seems impossible, a typo. It is not. What it means is that many children are infected with malaria more than once a year.

In North-Western Province, competent medical help can be difficult to find. For families living in the remote northern part of the province, across more than 1,000 square miles of wild terrain, there is only one place that can ensure a reasonable chance of survival when severe malaria strikes a child: Kalene Mission Hospital. This modest health center, in a decaying brick building capped with a rusty tin roof, represents the front line in the conflict between malaria and man. Scientists at the world's high-tech labs ponder the secrets of the parasite; aid agencies solicit donations; pharmaceutical companies organize drug trials. But it is Kalene Hospital — which functions with precisely one microscope, two registered nurses, occasional electricity from a diesel generator, and sometimes a doctor, sometimes not (though always with a good stock of antimalarial medicines) — that copes with malaria's victims.

Every year for a century, since Christian missionaries founded the hospital in 1906, the coming of the rainy season has marked the start of a desperate pilgrimage. Clouds gather; downpours erupt; mosquitoes hatch; malaria surges. There's no time to lose. Parents bundle up their sick children and make their way to Kalene Hospital.

They come mostly on foot. Some walk for days. They follow trails across borders, into rivers, through brushwood. When they reach the hospital, each child's name is printed on a card and filed in a worn wooden box at the nurses' station. Florence, Elijah, Ashili. They come through the heat and the rain and the dead dark of the cloudy night. Purity, Watson, Miniva. Some unconscious, some screaming, some locked in seizure. Nelson, Japhious, Kukena. A few families with bicycles, Chinese-made one-speeds, the father at the pedals, the mother on the seat, the child propped between. Delifia, Fideli, Sylvester. They fill up every bed in the children's ward, and they fill up the floor, and they fill up the courtyard. Methyline, Milton, Christine. They pour out of the bush, exhausted and dirty and panicked. They come to the hospital. And the battle for survival begins.

From the mosquito's salivary glands to the host's liver cell: a quiet trip. Everything seems fine. Even the liver itself, that reddish sack of blood-filtering cells, shows no sign of trouble. It's only in those few rooms whose locks have been picked by *falciparum* where all is pandemonium. Inside these cells, the malaria parasites eat and multiply. They do this nonstop for about a week, until the cell's original contents have been entirely digested and it is bulging with parasites like a soup can gone bad. Each *falciparum* that entered the body has now replicated itself 40,000 times.

The cells explode. A riot of parasites is set loose in the bloodstream. Within 30 seconds, though, the parasites have again entered the safe houses of cells — this time, each has drilled into a red blood cell flowing through the circulatory system. Over the next two days, the parasites continue to devour and proliferate stealthily. After they have consumed the invaded cells, they burst out again, and once more there is bedlam in the blood.

For the first time, the body realizes it has been ambushed. Headache and muscle pains are a sign that the immune system has been triggered. But if this is the victim's first bout of malaria, the immune response is mostly ineffective. The alarm has sounded, but the thieves are already under the bed: the parasites swiftly invade a new set of blood cells, and the sequence of reproduction and release continues.

Now the internal temperature begins to rise as the body attempts to cook away the invaders. Shivering sets in — muscle vibrations generate warmth. This is followed by severe fever, then drenching sweat. Cold, hot, wet; the symptoms are a hallmark of the disease. But the parasites' exponential growth continues, and after a few more cycles there are billions of them tumbling about the blood.

By this point, the fever has reached maximum intensity. The body is practically boiling itself to death — anything to halt the attack — but to no avail. The parasites can even commandeer blood cells to help aid their survival. In some cases of *falciparum*, infected cells sprout Velcro-like knobs on their surfaces, and as these cells pass through the capillaries of the brain, they latch on to the sides. The adhesion keeps them from washing into the spleen, which cleans the blood by shredding damaged cells. Somehow — no one is quite sure how — the adhesion also causes the brain to swell. The infection has turned into cerebral malaria, the most feared manifestation of the disease.

This is when the body starts to break down. The parasites have destroyed so many oxygen-carrying red cells that too few are left to sustain vital functions. The lungs fight for breath, and the heart struggles to pump. The blood acidifies. Brain cells die. The child struggles and convulses and finally falls into a coma.

Malaria is a confounding disease — often, it seems, contradictory to logic. Curing almost all malaria cases can be worse than curing none. Destroying fragile wetlands, in the world of malaria, is a noble act. Rachel Carson, the environmental icon, is a villain; her three-letter devil, DDT, is a savior. Carrying a gene for an excruciating and often fatal blood disorder, sickle-cell anemia, is a blessing, for it confers partial resistance to *falciparum*. Leading researchers at a hundred medical centers are working on antimalarial medicines, but a medicinal plant described 1,700 years ago may be the best remedy available. "In its ability to adapt and survive," says Robert Gwadz, who has studied malaria at the National Institutes of Health, near Washington, D.C., for almost thirty-five years, "the malaria parasite is a genius. It's smarter than we are."

The disease has been with humans since before we were human. Our hominin ancestors almost certainly suffered from malaria. The parasite and the mosquito are both ancient creatures — the dinosaurs might have had malaria — and this longevity has allowed the disease ample time to exploit the vulnerabilities of an immune system. And not just ours. Mice, birds, porcupines, lemurs, monkeys, and apes catch their own forms of malaria. Bats and snakes and flying squirrels have malaria.

Few civilizations, in all of history, have escaped the disease. Some Egyptian mummies have signs of malaria. Hippocrates documented the distinct stages of the illness; Alexander the Great likely died of it, leading to the unraveling of the Greek Empire. Malaria may have stopped the armies of both Attila the Hun and Genghis Khan.

The disease's name comes from the Italian *mal'aria*, meaning "bad air"; in Rome, where malaria raged for centuries, it was commonly believed that swamp fumes produced the illness. At least four popes died of it. It may have killed Dante, the Italian poet. George Washington suffered from malaria, as did Abraham Lincoln and Ulysses S. Grant. In the late 1800s, malaria was so bad in Washington, D.C., that one prominent physician lobbied — unsuccessfully — to erect a gigantic wire screen around the city. A million Union Army casualties in the U.S. Civil War are attributed to malaria, and in the Pacific theater of World War II casualties from the disease exceeded those from combat. Some scientists believe that one of every two people who have ever lived have died of malaria.

The first widely known remedy was discovered in present-day Peru and Ecuador. It was the bark of the cinchona tree, a close cousin of coffee. Local people called the remedy *quina quina* (bark of barks), and it was later distributed worldwide as quinine. Word of the medicine, spread by Jesuit missionaries, reached a malaria-ravaged Italy in 1632, and demand became overwhelming. Harvested by indigenous laborers and carried to the Pacific coast for shipment to Europe, the bark sold for a fortune.

Several expeditions were dispatched to bring seeds and saplings back to Europe. After arriving in South America, the quinine hunters endured a brutal trek through the snow-choked passes of the Andes and down into the cloud forests where the elusive tree grew. Many perished in the effort. And even if the quinine hunters didn't die, the plants almost always did. For two hundred years, until the cinchona tree was finally established on plantations in India, Sri Lanka, and Java, the only way to acquire the cure was directly from South America.

Quinine, which disrupts the malaria parasites' reproduction, has saved countless lives, but it has drawbacks. It is short-acting, and if taken too frequently can cause serious side effects, including hearing loss. In the 1940s, however, came the first of two extraordinary breakthroughs: a synthetic malaria medicine was introduced. The compound was named chloroquine, it was inexpensive and safe, and it afforded complete, long-lasting protection against all forms of malaria. In other words, it was a miracle.

The second innovation was equally miraculous. The Swiss chemist Paul Müller discovered the insecticidal power of a compound called dichlorodiphenyltrichloroethane, better known as DDT. Müller was awarded the 1948 Nobel Prize in medicine for his discovery, for nothing in the history of insect control had ever worked like DDT. Microscopic amounts could kill mosquitoes for months, long enough to disrupt the cycle of malaria transmission. It lasted twice as long as the next best insecticide and cost one-fourth as much.

Armed with the twin weapons of chloroquine and DDT, the

World Health Organization in 1955 launched the Global Malaria Eradication Programme. The goal was to eliminate the disease within ten years. More than a billion dollars was spent. Tens of thousands of tons of DDT were applied each year to control mosquitoes. India, where malaria had long been a plague, hired 150,000 workers, full-time, to spray homes. Chloroquine was widely distributed. It was probably the most elaborate international health initiative ever undertaken.

The campaign was inspired by early successes in Brazil and the United States. The United States had recorded millions of malaria cases during the 1930s, mostly in southern states. Then an intensive antimalaria program was launched. More than 3 million acres (1.2 million hectares) of wetlands were drained, DDT was sprayed in hundreds of thousands of homes, and in 1946 the Centers for Disease Control was founded in Atlanta specifically to combat malaria.

America's affluence was a major asset. Almost everyone could get to a doctor; windows could be screened; resources were available to bulldoze mosquito-breeding swamps. There's also the lucky fact that the country's two most common species of *Anopheles* mosquitoes prefer to feed on cattle rather than humans. By 1950 transmission of malaria was halted in the United States.

The global eradication effort did achieve some notable successes. Malaria was virtually wiped out in much of the Caribbean and South Pacific and in the Balkans and Taiwan. In Sri Lanka, there were 2.8 million cases of malaria in 1946 and a total of 17 in 1963. In India, malaria deaths plummeted from 800,000 a year to scarcely any.

But it was also clear that the campaign was far too ambitious. In much of the deep tropics malaria persisted stubbornly. Financing for the effort eventually withered, and the eradication program was abandoned in 1969. In many nations, this coincided with a decrease in foreign aid, with political instability and burgeoning poverty, and with overburdened public health services.

In several places where malaria had been on the brink of extinction, including both Sri Lanka and India, the disease came roaring back. And in much of sub-Saharan Africa, malaria eradication never really got started. The WHO program largely bypassed the continent, and smaller-scale efforts made little headway.

Soon after the program collapsed, mosquito control lost access to its crucial tool, DDT. The problem was overuse — not by malaria fighters but by farmers, especially cotton growers, trying to protect their crops. The spray was so cheap that doses many times stronger than necessary were sometimes applied. The insecticide accumulated in the soil and tainted watercourses. Though nontoxic to humans, DDT harmed peregrine falcons, sea lions, and salmon. In 1962 Rachel Carson published *Silent Spring*, documenting this abuse and painting so damning a picture that the chemical was eventually outlawed by most of the world for agricultural use. Exceptions were made for malaria control, but DDT became nearly impossible to procure. "The ban on DDT," says Gwadz of the National Institutes of Health, "may have killed 20 million children."

Then came the biggest crisis of all: widespread drug resistance. Malaria parasites reproduce so quickly that they evolve on fast-forward, constantly spinning out new mutations. Some mutations protected the parasites from chloroquine. The trait was swiftly passed to the next generation of parasites, and with each new exposure to chloroquine the drug-resistant parasites multiplied. Soon they were unleashing large-scale malaria epidemics for which treatment could be exceedingly difficult. By the 1990s, malaria afflicted a greater number of people, and was harder to cure, than ever.

The story of malaria is currently being written — by hand, in ballpoint pen — by the staff of Zambia's Kalene Mission Hospital. Every morning, soon after dawn, a nurse's aide who has just finished the night shift records a brief update on each child in the intensive care ward. The report is written on lined notebook paper and clipped into a weathered three-ring binder. The day workers add frequent notations on the small patient cards, kept at the nurses' station. Together, the night report and the cards form a compelling, immediate account of a deadly disease.

Many entries are simply terse, staccato jottings. "Mary: Has malaria. Unconscious." "Belinda: Malaria. Seizures." But others are far longer, enumerating clinical details about medicines and dosages and checkup times, as well as offering vivid glimpses into the struggle for survival in one of the world's most malarious places. Leaf through the pages, flip through the cards — there are thousands upon thousands of entries — and the stories emerge.

Here's Methyline Kumafumbo, a skinny three-year-old who was taken to Kalene Hospital by her grandmother. They journeyed ten miles from their home village, and by the time they arrived, malaria parasites had already latched on to Methyline's brain. "Admitted yesterday," the night report reads. "Fevers and seizures. Malaria." The right side of Methyline's head was shaved, and an IV line inserted. Quinine, which remains Kalene Hospital's frontline drug for severe cases, was administered, dose after dose, each treatment dutifully recorded.

For almost a week, Methyline languished in a coma. A malarial coma can be a horrible thing to observe: arched back, rigid arms, twisted hands, pointed toes. A still life of agony. The reports continue their unblinking assessment. "Unconscious. Continues on IV quinine." "Still unconscious though not seizuring." "Still unconscious."

Then the seizures started again. There are times when the night report reads almost like a personal diary. "I was worried," the aide wrote about Methyline. "So I informed Sister" — the honorific bestowed on the hospital's two nurses — "who came and ordered Valium, which was given with relief."

Finally, the entries turn hopeful. "She's opening up her eyes but she still looks cerebral." "Drinking and eating porridge." And then: "Is conscious and talking!!" Three days later, Methyline was released from the hospital. "Looking bright," says the report. "But still not walking well."

One insidious thing about malaria is that many who don't die end up scarred for life. "Her walking issues point to larger problems," Robert Gwadz says after reviewing the progression of Methyline's sickness. "She may have permanent neurological damage." This legacy of malaria has sobering repercussions for people and nations. "It's possible," says Gwadz, "that due to malaria, almost every child in Africa is in some way neurologically scarred."

And Methyline has to be considered one of the fortunate ones. The Kalene Hospital night report is filled with heartbreak. Christabel: "The patient is in bad condition. Grunting and weary. Irregular breathing. Sister was informed. Midnight she collapsed and died. The body was taken home. May her soul Rest in Peace." There's an entry like this on nearly every page. Ronaldo: "Semiconscious. IV for quinine. Seizure. Valium. Pain suppository. Fever.

More pain suppository. At 0500 hrs, child had gasping respiration. Finally, child suddenly collapsed and died. His body was taken home."

All of Zambia, it seems — from the army to the Boy Scouts to local theater troupes — has been mobilized to stop malaria. In 1985 the nation's malaria-control budget was $30,000. Now, supported with international grant money, it's more than $40 million. Posters have been hung throughout the country, informing people of malaria's causes and symptoms and stressing the importance of medical intervention. (The vast majority of the nation's malaria cases are never treated by professionals.) There are even Boy Scout merit badges for knowledge about malaria. Zambia's plan is to educate the public, then beat the disease through a three-pronged assault: drugs, sprays, and mosquito nets.

The country has dedicated itself to dispensing the newest malaria cure, which also happens to be based on one of the oldest: an herbal medicine derived from a weed related to sagebrush called sweet wormwood, or artemisia. This treatment was first described in a Chinese medical text written in the fourth century A.D. but seems to have been overlooked by the rest of the world until now. The new version, artemisinin, is as powerful as quinine with few of the side effects. It's the last remaining sure-fire malaria cure. Other drugs can still play a role in treatment, but the parasites have developed resistance to all of them, including quinine itself. To help reduce the odds that a mutation will also disarm artemisinin, derivatives of the drug are mixed with other compounds in an antimalarial barrage known as artemisinin-based combination therapy, or ACT.

Zambia is also purchasing enough insecticide to spray every house in several of the most malarious areas every year just before the rainy season. It has already returned to DDT — though just for indoor use, in controlled quantities. In the face of the growing malaria toll, access to DDT is gradually becoming easier, and even the Sierra Club does not oppose limited spraying for malaria control. Finally, the Zambian government is distributing insecticide-treated bed nets to ward off mosquitoes during the night, when the malaria-carrying *Anopheles* almost always bite.

The plan sounds straightforward, but progress against malaria never comes easily. Many Zambians living far from hospitals depend on roadside stalls for medicines. There ACT can cost more than a dollar a dose — virtually unaffordable in a country where more than 70 percent of the population survives on less than a dollar a day. So people buy other drugs for as little as 15 cents. They provide temporary relief, reducing the malarial fever, but may do little to halt the parasites.

Then there are widespread traditional beliefs. One of the posters plastered across Zambia reads: "Malaria is not transmitted by witchcraft, drinking dirty water, getting soaked in rain, or chewing immature sugar cane." When children suffer from seizures — a symptom of advanced cerebral malaria — some parents interpret it as a hex and head straight to a traditional healer. By the time they make it to the hospital, it's too late.

Even the gift of a bed net can backfire. There's no question that the nets can save lives, especially the latest types, which are impregnated with insecticide. But first they need to reach the people most in need, and then they must be properly used. "Distributing nets to remote villages is a nightmare," says Malama Muleba, executive director of the nonprofit Zambia Malaria Foundation. "It's one thing for me to convince Bill and Melinda Gates to donate money; it's quite another to actually get the nets out."

The Zambian army has been employed to help, but even after delivery, people can be reluctant to sleep beneath nets, which make a hot and stuffy part of the world feel hotter and stuffier. If a leg pops out at night or the fabric is torn, mosquitoes can still reach the skin. And the nets are sometimes misused as fishing gear. Theater troupes are spreading out into the Zambian countryside, emphasizing the proper use of bed nets through stage productions in settlements large and small.

Despite the difficulties, Zambia's campaign has started to produce results. In 2000 a study showed that fewer than 2 percent of children under the age of five slept under an insecticide-treated bed net. Six years later, the number had risen to 23 percent. The government of Zambia says an ACT known as Coartem is now available cost-free to the entire population. In a country that was steadily losing 50,000 children a year to malaria, early indications are that the death rate has already been reduced by more than a third.

But what if donor money dries up? What if Zambia's economy collapses? What about political instability? Both Angola and the Democratic Republic of the Congo, which flank Zambia, have a history of war. In the 1970s, during a civil war in Angola, six bombs landed near Kalene Mission Hospital; in the Congo war years, some of the nearby roads were mined.

"This is a critical moment," says Kent Campbell, program director of the Malaria Control and Evaluation Partnership in Africa. "There are no national models of success with malaria control in Africa. None. All we've seen is pessimism and failure. If Zambia is a success, it will have a domino effect. If it's a failure, donors will be discouraged and move on, and the problem will continue to get worse."

No matter how much time, money, and energy are expended on the effort, there still remains the most implacable of foes — biology itself. ACTs are potent, but malaria experts fear that resistance may eventually develop, depriving doctors of their best tool. Before the ban on DDT, there were already scattered reports of *Anopheles* mosquitoes resistant to the insecticide; with its return, there are sure to be more. Meanwhile, global warming may be allowing the insects to colonize higher altitudes and farther latitudes.

Drugs, sprays, and nets, it appears, will never be more than part of the solution. What's required is an even more decisive weapon. "When I look at the whole malaria situation," says Louis Miller, cochief of the malaria unit at the National Institute of Allergy and Infectious Diseases, "it all seems to come down to one basic idea: we sure need a vaccine."

It's easy to list every vaccine that can prevent a parasitic disease in humans. There are none. Vaccines exist for bacteria and viruses, but these are comparatively simple organisms. The polio virus, for example, consists of exactly 11 genes. *Plasmodium falciparum* has more than 5,000. It's this complexity, combined with the malaria parasite's constant motion — dodging like a fugitive from the mosquito to the human bloodstream to the liver to the red blood cells — that makes a vaccine fiendishly difficult to design.

Ideally, a malaria vaccine would provide lifelong protection. A lull in malaria transmission could cause many people to lose any immunity they have built up against the disease — even adults could, immunologically speaking, revert to infant status — rendering it more devastating if it returned. This is why a partial victory over malaria could be worse than total failure. *Falciparum* also has countless substrains (each river valley seems to have its own type), and a vaccine has to block them all. And of course the vaccine can leave no opening for the parasite to develop resistance. Creating a malaria vaccine is one of the most ambitious medical quests of all time.

Recent malaria history is fraught with grand pronouncements that turned out to be baseless. MALARIA VACCINE IS NEAR, announced a *New York Times* headline in 1984. "This is the last major hurdle," said one U.S. scientist quoted in the article. "There is no question now that we will have a vaccine. The rest is fine-tuning." Seven years of fine-tuning later, another *Times* headline summarized the result: EFFORT TO FIGHT MALARIA APPEARS TO HAVE FAILED. In the late 1990s, the Colombian immunologist Manuel Patarroyo claimed, with much media fanfare, that he had found the answer to malaria with his vaccine, SPf-66. Early results were tantalizing, but follow-up studies in Thailand showed it worked no better than a placebo.

At least ninety teams around the world are now working on some aspect of a vaccine; the British government, by way of incentive, has pledged to help purchase hundreds of millions of doses of any successful vaccine for donation to countries in need. The one closest to public release, developed by the pharmaceutical company GlaxoSmithKline Biologicals in collaboration with the U.S. Army, is called RTS,S. In a recent trial in Mozambique, it protected about half the inoculated children from severe malaria for more than a year.

Fifty percent isn't bad — RTS,S might save hundreds of thousands of lives — but it's not the magic bullet that would neutralize the disease once and for all. Many researchers suspect an all-encompassing cure isn't possible. Malaria has always afflicted us, they say, and always will. There is one man, however, who not only believes malaria can be defeated, he thinks he knows the key.

Stephen Hoffman is the founder and CEO of the only company in the world dedicated solely to finding a malaria vaccine. The company's name is Sanaria — that is, "healthy air," the opposite of malaria. Hoffman is fifty-eight, lean, and green-eyed, with a demeanor of single-minded intensity. "He's impassioned and impatient and intolerant of negativity," is how one colleague describes him.

Hoffman is intimately familiar with the pitfalls of the vaccine hunt. During his fourteen-year tenure as director of the malaria program at the Naval Medical Research Center, he was part of the team working on the vaccine promised in the 1984 *New York Times* article. He was so confident of the vaccine that he tested it on himself. He exposed himself to infected mosquitoes, then flew to a medical conference in California to deliver what he thought would be a triumphant presentation. The morning after he landed, he was already shaking and feverish — and, soon enough, suffering from full-blown malaria.

Now, more than two decades later, Hoffman is ready to return to prominence. He couldn't have found a more uninspiring launch pad: Sanaria is headquartered in a dismal mini-mall in suburban Maryland, near a picture-framing shop and a discount office-supply store. From the outside, there's no mention of the company's mission. A window badly in need of washing bears the company name in tiny adhesive letters. Hoffman realizes it's probably best if the office-supply customers aren't fully aware of what's going on a few doors away.

Inside, generating a hubbub of activity, are some thirty scientists from across the globe. The lab's centerpiece is a room where Hoffman raises mosquitoes infected with the *falciparum* parasite — yes, in a quiet mini-mall. Hoffman claims it's the world's most secure insectary. To enter, a visitor must pass through multiple antechambers sealed between sets of doors, like a lock system in a canal. Everyone has to wear white cotton overlayers, masks, shoe covers, and gloves. White makes it easier to see a stray mosquito. The air is recirculated, and the insectary is checked daily for leaks. Signs abound: WARNING! WARNING! INFECTIOUS AGENT IN USE. And hanging on a wall is a time-honored last line of defense: a fly swatter.

The mosquitoes are housed in a few dozen cylindrical containers, about the size of beach buckets, covered with mesh lids. They're fed *falciparum*-infected blood, then stored for two weeks while the parasites propagate in the insects' guts and migrate to the salivary glands, creating what are known as "loaded" mosquitoes. The loaded insects are transferred carefully to a kilnlike irradiator to be zapped with a quick dose of radiation. Then, in a special dissecting lab, the salivary glands of the mosquitoes are removed. Each mosquito's glands contain more than 100,000 parasites. Essentially, the vaccine consists of these irradiated parasites packed into a hypodermic needle. The idea is based on research done in the late 1960s at New York University by Ruth Nussenzweig, who demonstrated that parasites weakened by radiation can prompt an immune response in mice without causing malaria. Hoffman's vaccine will deliver the wallop of a thousand mosquito bites and, he says, produce a complete protective response. Thereafter, any time the vaccinated person is bitten by a malaria-carrying mosquito, the body, already in a state of alert, will not allow the disease to take hold.

Hoffman's lofty goal is to eventually immunize all 25 million infants born in sub-Saharan Africa every year. He believes that at least 90 percent of them will be protected completely from malaria. If so, they'll be the first generation of Africans, in all of human history, not to suffer from the disease.

But which generation will it be? Although Sanaria's vaccine may undergo initial field-testing next year, a federally approved version won't be available for at least five years — and maybe never. Given the track record of malaria vaccines, that's a distinct possibility. After so many million years on Earth and so many victories over humanity, the disease, it is certain, will not surrender easily.

When it comes to malaria, only one thing is guaranteed: every evening in the rainy season across much of the world, *Anopheles* mosquitoes will take wing, alert to the odors and warmth of living bodies. A female *Anopheles* needs to drink blood every three days. In a single feeding, which lasts as long as ten minutes, she can ingest about two and a half times her pre-meal weight — in human terms, the equivalent of downing a bathtub-size milk shake.

If she happens to feed on a person infected with malaria, parasites will accompany the blood. Two weeks later, when the mosquito flies through the open window of a mud hut, seeking her next meal, she'll be loaded.

Inside the hut, a child is sleeping with her sister and parents on a blanket spread over the floor. The family is aware of the malaria threat; they know of the rainy season's dangers. They've hung a bed net from the ceiling. But it's a steamy night, and the child has tossed and turned a few times before dropping back to sleep. Her foot is sticking out of the net. The mosquito senses it and dips down for a silent landing.

JAMES GEARY

The First Assassination of the Twenty-first Century

FROM *Popular Science*

IT BEGAN AS A STANDARD ADMISSION. When he arrived in the critical-care unit at University College Hospital (UCH) in central London on November 17, 2006, the patient in Room 9 was weak but alert. For just over two weeks he had been suffering from severe dehydration and vomiting. Comforted by a clutch of family and friends, he struggled to beat back an illness that was remorselessly attacking all his major organs. Physicians methodically disqualified the usual suspects: no food poisoning, no gastrointestinal infection. Then the patient's white-blood-cell count dropped to practically nothing and his hair began to fall out. He showed all the symptoms of acute radiation syndrome, but no radiation had been detected. "The Geiger-counter readings were negative," recalls Geoff Bellingan, the clinical director of the department of critical care at UCH. "There was no clarity on the diagnosis."

While the doctors struggled to identify his condition, the patient in Room 9 — Alexander Litvinenko, a vocal opponent of Russian president Vladimir Putin and an ex-officer in Russia's Federal Security Service (FSB), the successor to the KGB — had already reached his own conclusions. He was sure that he had been poisoned and that the Kremlin had ordered his assassination.

What started as an ordinary emergency-room case quickly blossomed into something larger: British researchers began to worry that this lone case might signify a health threat to the rest of the country. Litvinenko accused the Kremlin of seeking not only to kill him but to systematically wipe out its critics. And in the end, one man's murder became a glimpse into the future of assassination, a new world where high-tech hit men have access to terrible weapons.

Death of a Dissident

During the late 1990s, Alexander Litvinenko was assigned to the FSB's organized-crime unit. His job was to combat corruption in the aftermath of the country's chaotic transition to a free-market economy. But he became disillusioned with the security agency, and in 1998 he held a strange press conference at which he and several other disgruntled officers, some of whom wore ski masks to hide their identity, accused their bosses of seeking to line their own pockets and "settle accounts with undesirable persons." In 2000, after falling out with Putin, Litvinenko fled Russia for London — the destination of choice for Russia's restive dissidents and disaffected oligarchs — where he continued to antagonize his former colleagues.

Litvinenko claimed in a book, for instance, that the FSB was responsible for a series of apartment bombings in Russia in 1999. (The attacks were officially blamed on Chechen separatists, and Putin had used the incident to help justify a fresh invasion of Chechnya that same year.) He investigated the 2006 murder of the journalist and Putin critic Anna Politkovskaya. In February 2007, Alexander Gusak, Litvinenko's old commanding officer at the FSB, accused him of having revealed to British authorities the identities of Russian agents. "I was brought up on Soviet law," Gusak told the BBC's *Newsnight* television program. "That provides for the death penalty for treason. I think if in Soviet times he had come back to the USSR, [Litvinenko] would have been sentenced to death." A new law, adopted by the Russian parliament last year, authorizes the elimination of individuals outside of Russia whom the Kremlin accuses of terrorism or extremism. Litvinenko openly worried that his life was in danger. He was right.

His death began on November 1, 2006, when he met the FSB agents turned businessmen Dmitry Kovtun, Andrei Lugovoi, and, possibly, Vyacheslav Sokolenko for tea at London's Millennium Hotel. Later that night, he complained of vomiting, diarrhea, and fatigue. He checked into Barnet General Hospital in north London on November 3, but doctors couldn't find anything wrong. After exhausting the possibilities, and with Litvinenko's condition deteriorating, they transferred him to UCH on November 17.

Litvinenko's condition became progressively worse. Pictures of the former spy in his hospital bed show him looking paler than the crisp white walls in the UCH critical-care unit. On November 22, he was intubated and placed on mechanical ventilation. One by one, his vital organs — liver, kidneys, spleen — began to fail. His immune system collapsed as his white-blood-cell count plummeted. And still doctors did not know what was killing him. "We tried to examine his bone marrow, but it was so flat we couldn't get a sample," Bellingan says. "Something had poisoned all his dividing cells, but it wasn't clear which of many possible agents was involved."

Geiger counters failed to pick up any telltale gamma radiation — the easiest kind to detect — and radioactive thallium poisoning, an early hypothesis, had already been ruled out. "Once gamma was eliminated," Bellingan says, "we were looking at all comers. But the list of possible agents was very, very long." Litvinenko's case had distinguished itself as something new, something more complicated than the British medical establishment could handle. To pinpoint the poison, a sample of Litvinenko's urine went to Britain's Atomic Weapons Establishment (AWE). Researchers there detected signs of alpha radiation.

One of the strongest emitters of alpha radiation is an isotope called polonium-210, generally manufactured for industrial use in antistatic devices. The isotope quickly became the focus of the AWE investigation, but it was too late to do the patient any good. Back at UCH, Litvinenko was fading fast. "His heart was getting weaker and weaker," says Jim Down, the intensive-care consultant on duty the day Litvinenko died. "His blood pressure dropped inexorably to nothing." The AWE confirmed the polonium diagnosis at about 6 P.M. on November 23, but the news took several hours to reach the hospital. Before it did, at 9:21 P.M., the patient's heart gave out. He never knew the name of his poison.

Internal Decay

Nuclear physicists call polonium "the Terminator," not because of its efficacy as a poison but because it's the final element created in the process known as slow neutron capture. As an element, polonium occurs naturally in Earth's crust as a byproduct of the decay of uranium-238, and it accounts for about 1 percent of the total annual dose humans get from normal background radiation. In appearance it resembles a silvery gray dust — that is, if you can get enough of the stuff together to actually be able to see it. (Litvinenko received an amount that would have fit on the head of a pin with room to spare.) Polonium-210, created by bombarding bismuth-209 with neutrons inside a nuclear reactor, is hard to find in high concentrations. Only about 100 grams are produced every year, most of it in Russia.

Unlike many other radioactive substances, polonium-210 is harmless as long as it remains outside the body. Once inside the body, though, the alpha radiation emitted by the isotope is about twenty times as damaging to cells as the gamma radiation emitted by elements like thallium.

Gamma rays can penetrate steel, concrete, human tissue. Alpha particles can't penetrate even a single sheet of paper or your epidermis. But if you swallow an alpha emitter or inhale it, or if it enters the bloodstream through an open wound, all molecular hell breaks loose. "It's like firing a missile at a bag of Ping-Pong balls," says Paddy Regan, a lecturer in nuclear physics at the University of Surrey outside London. "If you coat the inside of a person's gut with alpha particles, the particles will kill every cell they come into contact with."

Police suspect that Litvinenko was poisoned at the Millennium Hotel on November 1 and that the polonium, most likely dissolved in some kind of tasteless liquid solution, was slipped into his tea before or during his meeting with the Russian businessmen. The polonium-210 lined Litvinenko's gastrointestinal tract. From there, it seeped into his bloodstream and spread throughout his body, first targeting rapidly dividing cells: hair, skin, stomach, bone marrow. He probably received a much larger dose than was strictly needed to kill him, somewhere between 1 and 10 gigabecquerels. (A becquerel is a measure of radioactivity amounting to one alpha-particle emission per second. Ten gigabecquerels, the maximum suspected dose, would have delivered 10 billion alpha emissions per second.) The amount was so great that he had no hope of survival. Litvinenko is the first person known to have died of polonium-210 exposure, and the first murdered with it.

The businessmen from the Millennium meeting deny any involvement in his death, although traces of radiation were found
along the paths they took in the days prior to the meeting. The trail
of alpha radiation across London suggests that whoever poisoned
Litvinenko did so at great personal risk. Inhaling it by accident, for
instance, would have meant certain death.

Sites within the hotel, as well as several items of tableware, showed
extremely high levels of polonium-210. The door to the men's
room was so contaminated that public health officials removed it
and disposed of it as nuclear waste. Litvinenko's home and office
were tainted by polonium-210, as were seats in airplanes, taxis, and
parts of a soccer stadium. You still can't book certain hotel rooms
in London because they're buzzing with traces of polonium-210.
"We're not dealing with scientists here who would have realized the
hazards of the material," says one source familiar with the investigation.

If the purpose of the assassination was to send a warning to other
dissidents, the assassins chose their weapon wisely: polonium-210
creates all the terror of a nuclear strike without the risk of mass fa-
talities. Since November 23, when polonium-210 was first identified in the case, the British Health Protection Agency has monitored about forty sites; at least twenty of them had significant levels
of polonium-210 contamination. The health agency has also tested
the urine of about seven hundred people, earning it the nickname
"the piss palace" among the staff. To date, seventeen individuals
have shown elevated levels of polonium-210.

Despite widespread fear of contamination, there was never any
threat to the general public. "Polonium is useless as a weapon of
mass destruction," Regan says. If it had been poured into London's
water supply instead of Litvinenko's tea, for example, it would have
dispersed so quickly that no one would have received a dangerous
dose.

And although they may not have understood the risks they were
taking, whoever masterminded the killing had clearly researched
their weapon to some extent. Nick Priest, a professor of radio-
biology at Middlesex University in England, estimates that it would
have taken a few days for a reactor to produce the amount of polo-
nium-210 delivered to Litvinenko. Polonium-210 has a half-life of
138 days, meaning that half of it will decay in about four and a half
months. So the assassins must have planned the operation well in

advance and then acted promptly. Also, the choice of polonium it-self suggests a certain sophistication. "They knew they could move it across borders because there is no gamma radiation," Priest says. "They knew that it would be taken up by gut. And they knew it was obscure. Even when doctors knew it was radiation [that was killing Litvinenko], they still didn't think of polonium." Now, of course, scientists and police think of little else.

Why, given all the methods available, did Litvinenko's killers choose polonium rather than a knife across the throat or a bullet to the head? "I think they supposed Litvinenko would die quickly," says Vladimir Ryzhkov, an independent member of the Russian parliament, "and that specialists wouldn't find out what substance was used. Polonium decays rapidly, so they may have expected no traces would be left behind and the British would say the cause of death was unknown."

In Russian political life, assassins who wish to remain anonymous often hide behind obscure methods. "You only need exotic ways of killing people when you don't want the truth to be revealed," says Alexei Kondaurov, a former KGB general who is now a parliamentarian critical of Putin's government. "But science has come a long way, and with modern methods of analysis, it's almost impossible to hide the truth."

Concern about contamination delayed Litvinenko's autopsy for a week, while officials discussed the precautions that had to be taken in cutting the dead man open. When the postmortem finally did take place, ventilation in the operating theater was switched off to prevent any wayward polonium-210 from becoming airborne, according to someone familiar with the procedure. Everyone stripped to his underwear before donning two separate impervious plastic suits, as well as a cylindrical shawl and helmet combination that slipped over the head and shoulders like a beekeeper's outfit. A filtration unit slung on a belt pumped scrubbed air into the suit. Tissue samples were passed through an airlock to a waiting pathologist, while a radiation-protection official continually monitored the room for alpha particles.

Wet Work

The former Soviet Union has always been one of the world's premier think tanks for exotic assassination methods. In the 1930s,

Stalin established a secret branch of the KGB with the fearsome name of the Administration for Special Tasks. The administration had a medical section called Kamera solely devoted to the development of exotic poisons and toxins. The "special tasks" this group administered consisted of what was known in the espionage jargon of the day as wet work: abducting and/or assassinating perceived "enemies of the people" wherever in the world they might be. One KGB memo stated, "As these traitors . . . have been sentenced to death in their absence, this sentence will be carried out abroad."

Radiation soon entered the arsenal for high-priority assassinations. In 1957, for instance, Nikolai Khokhlov — like Litvinenko a former agent turned critic who fled Russia to live in the West — took part in an anti-Soviet conference in Frankfurt, Germany. Shortly after sipping a cup of coffee that somebody handed him, he felt ill and fainted. Food poisoning was initially suspected, until strange lesions began appearing on Khokhlov's face and his hair started coming out in clumps. Doctors at an American military hospital eventually identified radioactive thallium and managed to save his life. Khokhlov felt confident that the Kremlin was behind the hit.

Now some regard last year's legislation authorizing killings outside of Russia — and a rash of recent assassinations — as a bit too reminiscent of the bad old days. "This is not a retreat to Soviet times," Kondaurov says, "but to one period of it, around 1937. [These killings show that] we're arriving at some violent authoritarian regime, a new quality of the Russian authorities that's similar to the worst examples of the past."

Kremlin officials have strenuously denied any involvement in Litvinenko's murder. And just because the polonium probably came from a Russian reactor doesn't mean that the assassination was officially sanctioned. "Reactors are making grams and grams of the material," says Middlesex University's Nick Priest. "It would not necessarily be noticed if a few micrograms went missing." Gennady Gudkov, a member of the Russian parliament and a former FSB officer, agrees. "There is no doubt that the people who killed Litvinenko are from Russia," he says. "There are no other leads in the Litvinenko case except Russian leads. But trying to connect the Russian leads with the state — these are very different things."

We may never know who orchestrated the Litvinenko murder,

but, like the radiation it left behind, the event has raised a frightening specter. The prospect of an increasingly authoritarian Russian regime, one that tucks vials of radioactive material into the breast pockets of hit men before dispatching them abroad to silence its critics, is certainly alarming. But even scarier, says Oksana Antonenko, the program director for Russia and Eurasia at the International Institute for Strategic Studies in London, is the prospect that the Kremlin had nothing to do with it.

Murder, Globalized

Since the collapse of the Soviet Union, the world has rightly feared what would happen if terrorists intent on mass murder managed to make a dirty bomb from one of the caches of nuclear material scattered across the region. The Litvinenko murder has created another anxiety entirely. In the pantheon of collective paranoia, weapons of mass destruction are now going to have to make room for the threat of targeted nuclear terrorism.

Putin's predilection for state control has one potentially positive effect in this regard: it suggests that he has a grip on nuclear safety and security. If, however, a rogue group in Russia obtained and deployed the polonium-210 on its own, it suggests that Putin's vaunted authority has limited reach. "Litvinenko's killing may be a sign that Putin is not as in control as the West believes he is," Antonenko says. "It may mean that nuclear material can still be acquired and that elements with access to it can still act independently."

The breakup of the Soviet Union and its massive military and nuclear infrastructure loosened control over a vast and frightening arsenal. Security has improved enormously since the chaos of the early 1990s, but there are still a large number of alarming sites — former biological-warfare laboratories, chemical-weapons facilities — throughout the country. Russia had an estimated 44,000 tons of biological agents (including plague, tularemia, anthrax, and smallpox) at the end of the Cold War. Just 20 percent were to be eliminated by the end of 2007. "The concern is that [these materials] might be susceptible to rogue elements, who could use them in pursuit of financial or other interests," Antonenko says.

It's hard to see what security services can do to stop traffic in substances that are dangerous even in microscopic doses, apart from installing ever more sophisticated detection technology and carrying out ever more invasive searches at vulnerable borders and transport hubs. And even extra measures like these won't always be successful. In January 2007, news emerged that 100 grams of highly enriched uranium had been seized the previous February in Tbilisi, the capital of Georgia. The smuggler tried to sell the material to an undercover police officer posing as a representative of a terrorist organization.

At the moment, of course, former Russian spies and current Russian dissidents, not everyday citizens, are the most likely targets for exotic assassinations using polonium or other unconventional weapons. But these threats are now part of the landscape, and there will always be someone prepared to spend the time and money necessary to use them. Litvinenko's murder is more than just a bizarre true-crime thriller. It's the first assassination of the twenty-first century, the first strike in a new world of high-tech murder.

Walter Litvinenko was in the room when his son passed away. "My son died," he said later, "and he was killed by a little nuclear bomb." The aftershocks from that explosion are still rippling through the world's security and intelligence communities. And perhaps that's just what the assassins intended. Maybe they wanted to send a message: this is a new and horrible way to die, and in the end, no one is safe from us.

ROBIN MARANTZ HENIG

Our Silver-Coated Future

FROM *OnEarth*

NANOTECHNOLOGY, fast becoming a $3-trillion industry, is about to revolutionize our world. Unfortunately, hardly anyone is stopping to ask whether it's safe. For an industry that trades in the very, very small, projections about the potential scope of nanotechnology are gigantic. Estimates are that the industry will grow at a staggering pace in its first decade, reaching close to $3 trillion globally by 2014. The National Nanotechnology Initiative, created by President Bill Clinton in 2000, has called it "the next industrial revolution." Enthusiasts say that nanotechnology may someday enable scientists to build objects from the atom up, leading to entirely new replacement parts for failing bodies and minds. It may enable engineers to make things that never existed before, creating nanosize "atomic carpenters" that can be programmed to construct anything, atom by atom — including themselves. Or it may make things disappear, with nanowires that get draped around an object in a way that makes the whole package invisible to the naked eye.

As difficult as it is to comprehend how huge the promise of nanotechnology is, it's just as hard to wrap your head around how tiny "nano" is. A nanometer is defined as 1 billionth of a meter, but what does that mean? The analogies are mind-boggling but not necessarily enlightening. Hearing how small things are when you're working at the nano level doesn't help you visualize anything exactly; all it does is make you sit back and say, "Wow." If you think of a meter as Earth, goes one analogy, then a nanometer would be a marble. If you think of a meter as the distance from Earth to the sun, then a nanometer would be the length of a football field. A nanometer is 1 hundred-thousandth the width of a human hair. Or, in a particularly kinetic description, it is the length that a man's beard will grow in the time it takes him to lift a razor to his face.

"Things get complex down there, in terms of the physics and the chemistry," says Andrew Maynard, chief science adviser for the Project on Emerging Nanotechnologies, established in 2005 at the Woodrow Wilson International Center for Scholars in Washington, D.C., in partnership with the Pew Charitable Trust. "When you have small blocks of stuff, they behave differently than when you have large blocks of stuff."

At the nano level, some compounds shift from inert to active, from electrical insulators to conductors, from fragile to tough. They can become stronger, lighter, more resilient. These transformed properties are what account for the infinite potential applications of nanoparticles, defined as anything less than about 100 nanometers in diameter.

The field is a textbook example of exponential growth. According to Lux Research, an emerging-technologies research and advisory firm based in New York, which has tracked the industry since 2001, the total value in 2004 of all products worldwide that incorporated nanotechnology was $13 billion. That figure grew to $32 billion in 2005 and to $50 billion in 2006, and Lux Research projects that it will reach $2.6 trillion by 2014.

Nanotechnology holds great potential for improving our lives. It might benefit the environment, for instance, by reducing our dependence on oil through the creation of a new power grid based on carbon nanotubes, which can carry up to a thousand times as much electricity as copper wiring without throwing off heat, and solar energy farms that use thin, cheap, flexible nano-engineered solar panels.

Nanostructures offer better options for rechargeable batteries, for instance, including the ones to be used in the next generation of hybrid cars. One such battery, made with nanostructured lithium–iron phosphate electrodes, is smaller, lighter, and less environmentally toxic than conventional lithium batteries and can hold more energy, take a charge more quickly, and maintain a charge longer, according to Michael Holman, a senior analyst with Lux Research. "It's not the compound itself that's nanoscale, but the surface of the material," Holman says. The surface of the battery electrode contains nanosize bumps and ridges, "which make the surface area much higher, allowing the electrons to flow in and out of it more quickly."

In the medical field, nanotechnology is expected to lead to dozens of innovations: new methods of cancer treatment that deliver chemotherapy directly to the tumor, earlier cancer detection using nanowires that can spot derangements in just a few protein cells, new methods of blood vessel grafting during heart surgery using nanoglue formed from nanospheres of silica coated in gold.

In cancer treatment, one application involves gold nanoshells: gold-coated glass spheres no more than 100 nanometers in diameter. These nanoshells enter tumors by slipping through tiny gaps in blood vessels that feed the malignancy. Once enough nanoshells accumulate in the tumor, scientists shine a near-infrared laser through the skin, heating up the gold particles and burning away the cancer. This technique, developed at the University of Texas Health Science Center, has worked in animal experiments and is about to be used in humans.

However, the real impact of nanotechnology, at least in the short term, will not be at the dramatic level of cancer cures or a new energy grid. For now, the technology will have to prove itself in the more mundane arena of commerce: washing machines that fight germs, antiseptic computer keyboards and kitchen utensils, windshields that repel rain, sunscreens that rub on easily and block the full spectrum of ultraviolet rays. Nanoparticles are being put into stain-resistant clothing (Haggar NanoTex pants with NANO-PEL), superlight tennis rackets (Wilson nCode), antiwrinkle face creams (Lancôme Rénergie Microlift), sunscreens (Blue Lizard), computer peripherals (IOGEAR), and a wall paint made by an Australian company, Nanovations, that says the paint can "achieve better energy ratings for buildings, better indoor air quality and fewer allergy-related illnesses."

But before we hurtle off toward a nano-utopia, we need to step back and ask ourselves whether this is a direction in which we really want to go.

When an industry grows this quickly, there may be neither the time nor the inclination to ask some tough questions about possible risks. First of all, there are the health and environmental hazards. Would nanotechnology bring unacceptable risks to workers making these materials or consumers who use the final products? Would it affect air or water quality near where the nanomaterials are dispersed? Very little is known about nanotoxicology, which might be very different from the toxicology of the same materials at normal scale.

Then there are the social, even existential, consequences. If the hype about nanotechnology contains even a smattering of truth, the technique could shake up our most basic assumptions about our place in the universe, turning us from its residents to the architects of its most fundamental elements. Might that act of hubris somehow subvert us as a species?

As nanotechnology explodes, and as federal agencies wrangle over whose responsibility it is to deal with an essentially unregulated industry, it's all the more crucial to take stock of the emerging field as soon as possible.

"This is not a technology we want to say no to out of hand," says Jennifer Sass, a senior scientist at the Natural Resources Defense Council (NRDC). "I think this is a technology that is potentially transformative, but we want to use it in a way to take advantage of that while reducing the risk."

Andrew Maynard sees this moment as a crossroads for nanotechnology. "What concerns me," he says, "is that if we're not smart about this we'll get something wrong, which would cause unnecessary damage to the environment or to people and would undermine the potential of all nanotechnology."

Nowhere is the tension between real and perceived risk — not to mention the tension between the mundane and the transformative — more apparent than with nanosilver. Nanosilver offers an important early test case for two reasons: it is now used in more consumer products than any other nanomaterial, and it is principally designed for use in products that come into direct contact with the human body. Since late 2005, the Project on Emerging Nanotechnologies has been compiling an inventory of products that contain nanomaterials. In the first two years the list more than doubled, to more than five hundred products as of the summer of 2007. Of these, nearly one hundred contain nanosilver, almost always because of its antimicrobial action.

In its ordinary form, silver is a metallic element with brilliant luster, great malleability, and the ability to conduct both heat and electricity. Its most common use is in photography, as silver halides and silver nitrate on photographic paper, and it has also long been used to make jewelry, coins, and tableware. Its luminosity has inspired songsmiths and poets; Emily Dickinson, for example, described the ocean as "an everywhere of silver."

Silver can also, in its regular form, kill bacteria, fungi, and other infectious microorganisms. (So can other metals, such as mercury and lead, but they, unlike silver, are almost as toxic to the human host as they are to microbes.) When silver is converted to nanosilver, this germ-killing quality is amplified, probably because of a change in surface-to-mass ratio. Silver's antimicrobial action is due to the release of positively charged silver ions on the surface, says Maynard, and "you get higher performance from the same mass of material" at the nanoscale. In addition, nanosilver can be incorporated into plastics, fabrics, and other consumer items more easily than can larger silver particles.

Nanosilver is added to socks and shoe liners to combat foot odor, to bandages to promote healing, to the insides of refrigerators and food storage containers to retard spoilage. It is applied to artificial joints and other implants to reduce the risk of infection. And there are nanosilver coatings or infusions in computer keyboards, computer mice, nail clippers, dog food bowls, spatulas, back support pillows, pay phones, air purifiers, handrails, ATM buttons — anywhere one set of hands might come into indirect contact with another set.

"But do we really need to put antimicrobial coatings on a computer mouse?" asks Maynard. "I mean, how many infections are really transmitted by someone using a mouse that has germs on it?"

Just before the 2006 flu season, the government of Hong Kong put nanosilver coatings on the handrails and grab-poles in the city's subway system to help prevent the spread of avian flu. The uncertainty over the health consequences of such an action was driven home in photographs of workers applying the nanosilver. As they sprayed on the nanomaterial — applied, remember, to protect the public's health — they were covered head to toe in protective hazmat suits. Granted, nanosilver might be more toxic in aerosol form than it is after it dries, but that's the point — no one knows for sure. The image of workers spraying stuff on handrails while wearing protective gear and face masks was, to say the least, disconcerting — and no one can say whether the nanosilver coating will remain intact once it dries or whether it will be rubbed off and dispersed after contact with thousands of commuters' hands.

And if hazmat suits are required for workers to apply nanosilver to handrails, what is nanosilver doing in Theramed S.O.S. Sensitive toothpaste? Or in a baby bottle made by the Korean company Baby Dream? Or in another child-care product from Korea, NANOVER Wet Wipes, which the manufacturer says are "soft like cotton, protect babies' frail skin"?

Silver has a long history of use in humans, and it has generally been found to be safe. In the days before antibiotics, silver was used as a curative; as long ago as the fourth century B.C., the Greek physician Hippocrates recommended as an ulcer treatment "the flowers of silver alone, in the finest powder."

Because of its germ-killing power, silver has taken on an almost mystical aura. According to folklore, silver repels vampires, and a silver bullet is the only way to kill a werewolf. Housewives in the early 1900s would drop a silver dollar into a bottle of milk, hoping to keep the milk fresh longer. And doctors of that era routinely administered eye drops of silver nitrate to newborns to prevent blindness that could result from an infant's exposure to gonorrhea, chlamydia, or other microorganisms living in the birth canal.

The alternative-medicine community has latched on to silver as an antimicrobial, too. Silver is sold in health food stores as colloidal silver, a liquid mixture of silver and water. A suspiciously broad range of claims has been made for colloidal silver, from healing wounds to treating skin cancer. Similarly extravagant claims are being made for nanosilver, as on the Web site of one distributor, Spirit of Ma'at of Sedona, Arizona, which states that its nanosilver supplement "protects against colds, flu, and hundreds of diseases (even anthrax)." The only health risk known to be associated with such supplements is an unsightly (though benign) condition known as argyria, in which the skin is permanently stained blue. (A Libertarian candidate in the 2006 U.S. Senate race in Montana, Stan Jones, who started using a homemade colloidal silver concoction in 1999, was famous for his ashen blue-gray skin.)

But might exposure to nanoscale particles of silver have more pernicious side effects? It's hard to say, because few studies have been done specifically on nanosilver. Despite this uncertainty, consumer products with nanosilver keep being introduced. And without any requirements for premarket safety testing, manufacturers have no incentive to conduct such tests on their own.

At the moment, the health risks of nanosilver are conjectural, based on what little is known about how other nanoparticles behave. But this is an imperfect system, since we can't be sure whether one nanoparticle's tendency to penetrate individual human cells predicts how a different nanoparticle — even a slightly different size or shape of the same basic nanomaterial — will behave.

What we know at this point is merely suggestive, but in some cases worrisome. One study of cells in culture, for instance, showed that when human lung tissue is exposed to carbon nanotubes, the lung cells see these not as foreign agents but as a biological substrate on which to build other tissue. Rather than mounting an immune response to attack the nanotubes as invaders, the lung cells start building layers of collagen around them. No one can say how likely it is in real life that carbon nanotubes would be inhaled; for most current uses, such as lightweight bicycle parts or tennis rackets, they are fixed in a matrix. But there is a chance that they might be inhaled during manufacture or as the product degrades, either through normal wear and tear or after it's disposed of. If they get into people's lungs, will carbon nanotubes act in vivo the way they do in cell culture and become a scaffolding for new layers of collagen that could block the airway?

Similar questions about the safety of nanoparticles arise from animal models showing that they can get into the bloodstream through the skin and then travel to vital organs, including the brain. As with airborne exposure, the likelihood of skin exposure to carbon nanotubes is still unknown, but once again early research indicates that there could be some health effects. Toxicologist Günter Oberdörster of the University of Rochester, working with rodents, found that carbon nanoparticles were small enough to enter the brain by way of the olfactory nerve, circumventing the blood-brain barrier, the usually impermeable membrane that protects the brain from foreign agents.

The main thing that is known about the toxicology of nanoparticles is how much remains to be discovered. Nanoparticles, says Maynard, "can penetrate into cells in ways that larger particles cannot or migrate to places in the body large particles cannot get to."

Another worrisome finding is a possible link between nanoparticles and the more rapid formation of protein fibrils, a material found in neurons that, when it accumulates, can lead to the buildup of a brain toxin called amyloid. Chemists from several European universities, led by Sara Linse of University College Dublin, exposed a laboratory preparation of purified protein to four types of nanoparticles, including carbon nanotubes and so-called quantum dots (crystals just 5 or 10 nanometers in diameter that are used in the semiconductor industry to measure electric current down to the level of the electron). All four types of nanoparticles accelerated the abnormal development of the protein into amyloid fibrils. The reason for the concern is that amyloid has been implicated in a variety of neurological diseases, including Alzheimer's and Parkinson's.

As with most studies in nanotoxicology, the Linse study is preliminary; it was conducted in vitro, not in an animal or human, and it remains to be seen whether the findings will be replicated in vivo. But it does point out the complexity of the emerging field of nanotoxicology.

"One of the most important messages of this work for chemists," wrote Vicki Colvin and Kristen Kulinowski of Rice University in the May 2007 *Proceedings of the National Academy of Sciences,* "is that when NP's [nanoparticles] enter the biological world they become very different materials." According to Colvin and Kulinowski, "The small sizes of NP's convey the potential to access many biological compartments, where they are met with a smorgasbord of possible binding partners from the complex and concentrated soup of biomolecules."

In terms of environmental consequences, if nanosilver is anything like ordinary silver, we might be in for some trouble. As with the potential human health risks, the environmental dangers can only be guessed at by analogy — in this case, to the known impact of normal-scale silver on aquatic organisms. According to Samuel Luoma, a senior research scientist at the U.S. Geological Survey in Menlo Park, California, silver is a powerful environmental toxin, second only to mercury in the damage to invertebrates that even trace amounts can do. It kills microorganisms indiscriminately and can wipe out the beneficial ones as well as the pathogenic ones. In addition, it has a direct effect on the reproductive capabilities of certain aquatic invertebrates and possibly fish as well.

Through most of the 1980s, says Luoma, silver pollution from a photo-processing plant led to widespread sterility among the *Macoma balthica* clams in South San Francisco Bay near Palo Alto. Clams are an important part of the bay-bottom food web, he says, and the population recovered only when new regulations limited the amount of silver in the bay. The photo-processing plant was eventually closed.

The lesson learned was a crucial one: it took a very low concentration of silver, less than one part per billion, to destroy the reproductive organs of virtually all adult *M. balthica* living within a certain distance of the silver source and to spread silver contamination throughout South Bay.

"No one outfall [the sewage pipe that carries wastewater from a treatment plant] could have effects everywhere in San Francisco Bay," Luoma says. "The risk with nanosilver from consumer products is that it would come from all outfalls that serve urban customers. What we need to know is how much silver would be released from many households. Would it be comparable to the mass from photo processing?"

Silver might have been especially toxic in San Francisco Bay because the bay is salt water. Unlike many other environmental toxins, silver seems to be more dangerous in salt water than in fresh. In fresh water, silver combines with chloride and forms a solid that sinks to the bottom, becoming less bioavailable (that is, less capable of absorption by the body). But in salt water, with a preponderance of chloride atoms, more silver chloride remains in solution, binding to particulate matter. This puts more of it into the food chain and therefore is more likely to do damage to marine organisms.

No other metal has this property of behaving one way in fresh water and another way in salt water. This presents a regulatory problem that's unique to silver, especially at the nano scale. According to Luoma, most toxicity testing is done in fresh water. "But a silver nanoparticle could look innocuous in fresh water and be extremely toxic in seawater," he says. How significant is this? Nobody really knows — but Luoma is concerned. It's reasonable, he says, to expect that nanosilver will shed from treated fabrics and from the linings of washing machines and food containers and make its way into rivers and streams, eventually ending up in the ocean.

There is also some evidence that excessive use of silver as an antimicrobial can lead to silver resistance in bacteria, in much the same way that excessive use of antibiotics can lead to the development of antibiotic-resistant organisms. In the late 1970s, for instance, scientists grew a laboratory culture of the common bacteria *E. coli* and exposed it to low levels of silver; within a few generations, the *E. coli* developed resistance to silver. If *E. coli* could do it, could other, potentially more dangerous microbes do the same thing? And if it happened with normal-scale silver, would it be more or less likely to happen with nanosilver? At the very least, nanosilver complicates the picture, since it allows silver to be used in so many more products. "If we use it too widely," Maynard says, "we may be giving away our best weapon."

Are we foolish to forge ahead in developing nanosilver products without full toxicology information? Perhaps. Luoma, who lives in Silicon Valley, says that the frenzy surrounding nanotechnology, the rush to be first at any cost, reminds him of the heedless gold-rush mentality of the dot-com era. A lot of nano-promises might fail to materialize, as happened with so many brilliant Internet startup ideas. The crucial difference is that the dot-com boom did no harm to the environment or to human health while the Darwinian struggle for survival played itself out. Nanotechnology might.

Here's how one product made from nanosilver, a set of kitchen utensils available in the United States, is being promoted by its manufacturer, Nano Care Technology of Hong Kong: "People always use traditional ways such as sterilizer to kill bacteria and germs but the result is not satisfied [*sic*], because many bacteria and viruses survive or relive [*sic*] very quickly." But the company's nanosilver kitchen utensils may do the job permanently, its website continues, and "can prevent people from the following diseases: duodenitis caused by spirillums, virosis hepatitis, dysentery caused by salmonella and food poisoning caused by golden staphylococcus."

The Korean appliance manufacturer Daewoo makes similar claims for its products treated with nanosilver (currently distributed only in Europe), which include a washing machine, refrigerator, and vacuum cleaner. It's clear from the Daewoo Web site that the company is using nanosilver for its antimicrobial properties: "After splitting the particles of silver known to have superior deodorant and antibiotic power by 1/1,000,000 mm, we have applied it to major parts of [the] refrigerator in order to restrain the growth and increase of a wide variety of bacteria and eliminate odor particles." Not only is nanosilver a disinfectant and deodorant, the company writes, in an English-language translation so elliptical as to make the true meaning unclear, it also "maintains balance of hormone [*sic*] within our body and intercepts electromagnetic waves significantly."

At the moment, claims like Nano Care's and Daewoo's exist in a regulatory limbo. No single agency has jurisdiction over nanomaterials (the same applies to many materials of conventional size); it depends largely on how a product is used or where it is in its life cycle. During its manufacture, a nanoparticle might fall under the jurisdiction of the Occupational Safety and Health Administration, which deals with workplace exposure. After that, if it is to be ingested or used in a drug or medical device, it might be regulated by the Food and Drug Administration. Once it's discarded, it might fall under the purview of the Environmental Protection Agency (EPA), charged with minimizing air- and waterborne toxins.

Federal agencies have been turning their backs on regulating nanotechnology, according to a report issued in May 2007 by the Wilson Center, largely because they are not convinced it warrants anything beyond the regulations already in place for standard-scale chemicals. But this might be a dangerous assumption, writes J. Clarence Davies, a senior fellow at Resources for the Future, a nonpartisan research center, in "EPA and Nanotechnology: Oversight for the 21st Century." According to Davies, "The relationship between science and regulation is complex" and filled with uncertainties. The best course is therefore "striking a balance between the harm that could be done by proceeding with an innovation and the harm that could be done by not proceeding."

Nanosilver is something of a jurisdictional oddity. Initially the EPA decided that a nanosilver-releasing Samsung washing machine was a device, like a fly swatter, and not a pesticide. But after public pressure from several interest groups, including the NRDC, the agency reversed itself. It placed nanosilver — an antimicrobial agent — under the authority of the Federal Insecticide, Fungicide, and Rodenticide Act (FIFRA) rather than the Toxic Substance Control Act (TSCA), under which most other chemicals are regulated. FIFRA requires manufacturers to submit toxicity data before a product can be approved for sale and gives EPA broad authority to prohibit or limit the sale of pesticides.

"There is something surreal about asking whether washing machines or food-storage containers are pesticides," notes Davies in his report, "and it is a type of problem not envisioned by the drafters of the FIFRA statute." His reading of the act is that what matters in classifying an ingredient as a pesticide is not so much what the manufacturer says it does as what the ingredient has been put there to do. "It's less a matter of claim than intent," he says. "And in my opinion, it's pretty easy to show that the silver isn't doing anything in the products we're talking about other than acting as a pesticide."

After EPA's initial FIFRA decision, however, the agency decided to regulate only specific nanosilver products — those that make explicit claims of antibacterial action. The result has been that several manufacturers have changed their claims from "kills germs" to less obvious formulations, such as "specially patented" or "stays fresh longer."

In early 2006, for instance, The Sharper Image was saying that its FresherLonger food-storage containers were "infused with naturally antibacterial silver nanoparticles." The company's website at the time featured photographs of strawberries and grapes stored for eight days in FresherLonger, compared to fruit stored for the same period in a conventional container. The conventionally stored fruit had grown "furry," the company wrote, and the FresherLonger fruit looked almost as good as new. The difference? "The silver nanoparticle miracle," according to Web site archives from April 2006. "In tests comparing FresherLonger to conventional containers, the 24-hour growth of bacteria inside FresherLonger containers was reduced by over 98 percent because of the silver nanoparticles!"

Berries stayed fresh, according to the 2006 website, because "patent-pending antimicrobial silver nanoparticles infused into the containers reduce growth of mold and fungus . . . Silver in microscopic particle form is a safe, medically proven antibacterial agent. That is why silver nanoparticles are infused into the polypropylene containers of the FresherLonger system."

By May 2007 the text and the fruit photos were gone from the website, replaced with a bland description of the product as a "specially treated" polypropylene container that "helps to retard spoilage." Nothing nano is even mentioned. All that is specified is a "patent pending" airtight seal, and the fact that the containers are durable, dishwasher safe, and translucent, so you can see what's stored inside.

The Sharper Image also makes slippers with nanosilver, but while they were once called "Contour-Foam Silver Slippers" — a name that had a kind of Wizard of Oz ring to it — today the slippers are described simply as "Contour-Foam." The company's website in April 2006 said they were made with "viscoelastic foam insoles infused with microscopic particles of silver that is naturally antibacterial and reduces growth of odor-causing bacteria." The Web site today makes no mention of silver or of bacteria fighting.

A company spokesman declined to comment when asked to explain The Sharper Image's marketing decisions. But Patrick Lin, director of the Nanoethics Group, in San Luis Obispo, California, offers one possible explanation: manufacturers are trying to have their cake and eat it too. "The manufacturers say that nanosilver is the key ingredient to kill bacteria in your laundry," he wrote in an e-mail follow-up to a telephone interview, "but in the same breath, they say (at least implicitly) that nanosilver won't have any significant impact after [it is] released into the water system. Well, which is it — is nanosilver an effective killer or not?"

The new science of nanotechnology is poised precariously between two vistas. In one direction, scientific researchers and industries scramble to capitalize on the technique's alluring potential; in the other, regulatory agencies and environmental groups debate ways of keeping risks to a minimum. Complicating the tasks in both directions, the exploitation and the regulation, is something that can be thought of as Nano's Paradox: the qualities that make nanoparticles a potential threat to health and the environment are the very qualities that offer a wonderful opportunity to improve that same health and environment.

Despite the uncertainty, even nanotechnology's critics stop short of calling for a moratorium. "Testing can be done on individual nano materials and products," Davies writes in "EPA and Nanotechnology," "and judgments on limiting production or marketing should be based on the results of these tests."

But just because testing can be done does not mean it will be done. Nor does it mean that scientists are even sure exactly how toxicology testing for nanoparticles should proceed. And the political process of imposing new regulations grinds slowly — much more slowly than the growth of the field of nanotechnology. This is why Maynard, for one, urges that we think in terms of "oversight" rather than regulation. "If you're looking at developing best practices for handling nanomaterials," he says, "you can be far faster than you can with new legislation leading to regulation. So there are ways of dealing with challenges in the near future that don't necessarily mean resorting to regulation."

When Maynard refers to oversight, he means whatever works to monitor and manage the impact of nanotechnology. Government regulation is just one tool for this kind of management; others, he says, include stewardship programs developed by industry, voluntary programs pushed by government, and guidance on safe practices in manufacture and disposal.

To Davies, who was one of the original architects of the EPA back in 1970, nanotechnology offers a chance to rethink the government's creaky regulatory apparatus altogether. He urges an overhaul of the EPA; a joint government-industry research institute on nanotoxicology; an interagency regulatory coordinating group, coupled with oversight committees in the House and Senate, focused on nanotechnology; and an annual appropriation of $50 million specifically for EPA research into the potential risk of nanomaterials to individual health and the environment ($38 million is currently spent on this by the federal government as a whole). Agency officials did not respond to repeated requests for comment on Davies's proposals.

"From a scientific perspective the field is incredibly exciting," says Sass of the NRDC. "From a regulatory perspective, I sympathize — it's a quagmire. But the real problem is from the economic perspective. Nanotechnologies are already out there in the marketplace, and we can't keep putting this stuff in products until we know more."

EDWARD HOAGLAND

Children Are Diamonds

FROM *Portland*

IN AFRICA, EVERYTHING IS AN EMERGENCY. Your radiator blows out, and as you solder a repair job, kids emerge from the bush, belonging to a village that you'll never see and reachable by a path you hadn't noticed. Though one of them has a Kalashnikov, they aren't threatening, only hungry. Eight or ten of them, aged eight or ten, they don't expect to be fed by you or any other strange adult. Although you know some Swahili, you can't converse, not knowing Lango, but because there is plenty of water in the streams roundabout, they are fascinated that you choose to drink instead from bottles you have brought. Gradually growing bold enough to peer into the open windows of your truck, they don't attempt to fiddle with the door or reach inside, seeing no food or curious mechanical delectables. The boxes packed there white-man-style are cryptically uninformative. Meningitis and polio vaccines, malaria meds, deworming pills, folic acid, vitamin A, and similar famine fighters. However, the kids will remain as long as you do, and you don't dare leave because this fabric of politesse would tear if you did, as it would have already if they were five years older. You wish you could ask them if mines have been laid in the road recently by either the rebels or the government forces. Their fathers, the men of the village, haven't emerged because they're probably off with the guerrillas, and the women would not show themselves in time of war anyway. It's a balance you must maintain here: friendliness and mystery.

The city splits at the seams with squatter camps, swollen by an enormous flux of displaced refugees from within hungry Kenya itself,

not to mention all the illegals from the civil wars afire in the countries that surround it: Somalia, Sudan, Ethiopia, Congo, Rwanda.
Look on a map: dire suffering. Need I say more?

The joke, if you can call it that, among whites here is if you feel a
hand grope for your wallet, the second thing to do is try to save the
life of the pickpocket. This is a city veering into calamity, where
transient whites like me still dribble in because it's a hub for aid
groups, and yet it's a traditional wash-up spot for Anglo ne'er-do-
wells who try to define themselves by where they have been. With
the AIDS pandemic, it will soon be too late for a number of things.

I help out in a place where we feed street kids and treat them with
skin ointments, antibiotics, inoculations, minerals, vitamins, whatever we happen to have. Powdered milk, powdered eggs, surplus
soups or porridges that another nongovernmental organization
may have given us. We have a basketball hoop up, and soccer balls,
board games, playing cards, a tent fly hooked to the back wall in
the courtyard with cots arranged underneath it as a shelter where
the children can feel some safety in numbers at least. What makes
you burn out are the ones dying visibly of AIDS. Yet you don't want
to banish them again to the furnace of the streets or, on the other
hand, specialize merely as a hospice, where salvageable kids aren't
going to want to come. Many of them wish to go to school but
have no home to go to school from or money for the fees. So I
scrounged a blackboard and teach addition, subtraction, geography, the English alphabet, when I have a break from refereeing a
gritty soccer game or supervising the dishwashing or triaging kids
with fevers or contusions who ought to go to the hospital (not that
that Dickensian trip is often in their best interest). We have artful
dodgers eating our fruits and sandwiches between excursions into
robbery, drugs, peddling — but also earnest tykes, plenty of them,
whom your heart absolutely goes out to. Yet triage is frustrating.

You don't have to be a doctor to help people who have no aspi-
rin or disinfectant or pills for malaria, tuberculosis, dysentery, or
epilepsy, no splints or bandaging, and no other nearby facility to
walk to in the bush. No Kaopectate, cough suppressants, malnutrition supplements, antibiotics for bilharziasis or sleeping sickness or
yaws. If you were a nurse, patients would be brought to you with

these diseases or hepatitis or broken limbs. The old stone and concrete ruins of a Catholic chapel that has been forgotten since the
colonial powers left could be reoccupied if you chased the leopards
and the cobras out, because joy is what is partly needed, especially
at first, and joy, I think, is, like photosynthesis for plants, an evidence of God. But joy, like beauty, is a continuum too, and in tem-
perate climates it waxes with the sun somewhat as plants do.

I can do the basic mechanics if we break down on the road, and
I know when to speed up or — equally important — slow down,
when figures with guns appear to block our passage. (If it's soldiers,
you never speed up, but the decision is not that easy because every
male can look like a soldier in a war zone, and the soldiers look like
civilians.) The big groups, such as Doctors Without Borders, CARE,
Oxfam, Save the Children, have salaried international staff they
can fly in from Honduras, Bangkok, or New Delhi to plug a momentary defection or a flip-out — dedicated career people, like
the UN's ladies and gentlemen, with New York, Geneva, London,
Paris, Rome behind them, who've been vetted: not much fooling
around. But there are various smaller outfits, whose fliers you don't
receive in the mail back home, that will hire "the spiritual drifter,"
as my friend Al put it to me, to haul pallets of plywood, bags of ce-
ment, first-aid kits in bulk, and bags of potatoes, bayou rice, cases of
your basic tins, like corned beef, tuna, salmon, peas, what-have-you,
and trunks of medicine to provision the solo picayune apostle out
doing Christ's appalling work in the hinterlands.

   You draw up lists of refugees so no one gets double their ration
by coming through the corn queue twice. Use ink stamped on
the wrists if you have to — if their names are always Mohammed or
Josephine. And you census the children as well and weigh a sampling of them in a sling scale, plus measure their upper-arm fat, if
they have any, with calipers to compile the ratio of malnutrition in
the populace, severe versus moderate, and so on. I've helped inject
against measles, tetanus, typhoid, when not enough licensed people were there, having been a vet's assistant at one point in my
teens. I've powwowed with the traditional clan chiefs and tribal
healers, the leopard-skin priests and village shamans and elders, or
young militia commanders, and delivered babies when nobody
competent was around. I've squeezed the rehydration salts into babies' mouths when they're at death's door, mixed the fortified for-

mula that you spoon into them, and chalked the rows of little white squares in the dirt where you have all the children sit individually at their feeding hours so that every individual gets the same amount of protein, the same units of vitamins A, C, E, B, calcium, iron, phosphorus, out of the 55-gallon steel drum you're stewing the emergency preparation in. *Hundreds* of passive, dying children sitting cross-legged in the little squares, waiting for you to reach each of them. You don't think that breaks your heart? Chalk is never gonna look the same.

You meet many travelers: businessmen with attaché cases full of bank notes to persuade the bureaucrats in Government House to sign on to a certain project scheme; ecologists on a mission to save the chimpanzees; trust-fund hippies doing this route overland, now that you can't go from Istanbul into Afghanistan; specialists from one of the UN's many agencies studying a development proposal or transiting to the more difficult terrain of Rwanda, Zimbabwe, Somalia, then resting for a spell on the way back. The taxi stand is busy from sunrise to pitch dark, and the pool on the roof is patronized by African middle-class parents, some of whom are teaching their kids how to swim, as well as the KLM airline pilots and Swissair stewardesses, the Danish or USAID water-project administrators waiting for permanent housing, or bustling missionaries passing through.

Food is so central that you can't exaggerate the issue. My waitress, for example, was hungry for protein even though she worked in an expatriates' hotel. There was a pot of gruel in the kitchen for the help, but it wasn't nourishing enough for a lactating mother, and the chicken parts and fresh fruit, meat, and vegetables the guests ate were exactly inventoried each night to compare what remained in the refrigerator with the restaurant's orders. Nor did she eat leftovers from people's plates, because she'd heard that AIDS can be spread by saliva. And, although she wouldn't be robbed at night once she reached her bus station, she said that to get to it she needed to use her tips to take a taxi because men with clubs waited next to a corner between here and there.

The fabled "Cape to Cairo" artery from South Africa toward Khartoum and Egypt is no more — a jaunt that was throttled first by Idi

Amin and is now choked off again by the war in southern Sudan. Beyond the city's suburbs, indeed, signs of human occupancy almost disappear in the elephant grass and regrown jungle, because for the next eighty or a hundred miles all of this has been a triangle of death, incinerated in Uganda's own civil wars: Amin's eight years of butcheries and then, when he was overthrown, Obote, who in an ostensibly saner manner killed just as many until Museveni upended him. So many clanking tanks, half-tracks, and fearsome assault platoons crawled up this road, mowing down anything that moved, that even in the peace Museveni established, nobody wanted to take the chance of living anywhere that might be visible in the forest. Bower birds, sunbirds, bee-eaters, secretary birds, reedbucks, dik-diks, spitting cobras, and black mambas were about, but the people who had somehow survived on site or returned afterward were not going to trust the neighborhood of the road, except for bicycling quickly along the asphalt. And then the trick was to vanish imperceptibly when you reached your destination, with no path to a hidden village that soldiers rumbling by in a grisly truck would notice. Settlements had been scorched, torched, eviscerated horribly, the skulls piled up from the massacres. And women still sometimes flinched — broke into a run for the woods — at the sound of my motor.

In the sixties, after Uganda's independence from the British, it had been mainly these local forest tribes, such as the Baris and Acholis, who logically should have been part of Uganda to begin with anyway, fighting the Muslim government. Then in the seventies a new president of Sudan had made peace with them, with regional freedom of religion and cultural autonomy, until the eighties, when a new fundamentalist Sharia swing reignited the war, led now by the big plains tribes, Nuer and Dinka, pastoralists living between these mountains and forests and the vast Sahel of the Arabs. It was a more serious insurrection, with the southern black army officers defecting to command the rebels, and, on the government side, the Baggara Arab tribes armed for lethal, devastating cattle raids against their neighbors the Dinkas. The United States was allied with Khartoum for strategic reasons at the time — to harass Muammar al-Qaddafi on Libya's flank — and discouraged even food aid reaching the southerners. So a quarter of a million people starved.

*

One morning in Sudan I woke to a lovely morning sun and bound-
less rolling perspectives after I climbed to a viewpoint behind the
church that overlooked the modest gorge of the Bahr el Jebel, the
"Mountain Nile." It flows under the Imatong Range, Sudan's highest, its ridges behind me, all downstream from the Victoria and Albert sections of the Nile in Uganda's lake country. Near Malakal it
is joined by the Bahr el Ghazal, the "Gazelle River," from the west,
and the Sobat, from the east, to form the White Nile. Then at Khartoum, the Blue Nile, from Ethiopia, joins the White to constitute
the famous Nile that flows to Shendi, Atbara, Wadi Haifa, Aswan,
Luxor, Cairo, and the delta close to Alexandria: nowhere, though,
is it more beautiful than around here. Beyond the gorge sit endless
savanna grasslands, woodlands, parkland, in tropical, light-filled
yellows and greens, where, although the hartebeests, kobs, buffalo,
and reedbucks may already have been eaten and the rhinos and elephants shot for their horns or tusks to buy guns, the vistas remain
primeval because for decades civil war has prevented any other
kind of development, like logging, tourism, or mining. I had a
spear-length stick in hand to defend myself afoot or keep the wildlife at bay and, more appropriately, to remind myself of how the
Dinkas, as a cattle people from time immemorial, had been able to
protect their herds and pasturage from the Baggara Arab tribes
whose homelands adjoined theirs, even though the Baggara domesticated horses as well as cattle and rode into battle, instead of
merely running. A Dinka, who could run for twenty miles with six
spears in his free hand, attacking from the reeds and rushes of
every river crossing, every hyacinth swamp, was not a foe whose cattle could be rustled and women stolen with impunity. But the equilibrium of spear versus spear had been skewed when Khartoum
gave the Baggara guns and sent its army in motor vehicles and helicopter gunships to mow down the lumbering cattle who escaped
the horses, driving the surviving herdsmen off their beloved prai-
ries, steppes, swamps, and plains.

We were inoculating babies against measles before the dry ice that
kept the vaccine fresh was gone. But an elephantiasis sufferer was
in the line. How could she have survived this long? We gave her
Cephalexin, for whatever that might be worth. The queue was
checkered with people with rag-fashioned bandaging, wearing blue
robes for one particular clan, red for another, with body paint,

headcloths, loincloths, tribal scarifications on the forehead. One man had what was possibly a giraffe's scrotum as a carryall on a rawhide strap from his shoulder, yet he was wearing a garage-sale apron from Peoria.

As we drove north, the people walking were sparse, and mostly from the local Madi or Bari tribes, a foot shorter than the Dinkas and thicker-set, scrambling along like forest-and-mountain folk, not striding like cattle herders, plainsmen. As food got shorter again the next week, they would disappear into the woods, and gaunt Dinkas would take over the roads, famished, stalking something to eat. We saw a burly man with a bushbuck that he had snared slung over his shoulder, who started to run when he heard the motor, till he realized we were aid workers and not about to steal his meat, as guys in a Sudan People's Liberation Army vehicle would have. Another man, two miles on, was squatting on his heels, quietly collecting wood doves one by one every few minutes, when they fluttered down to drink at a roadside puddle he had poisoned with the juice of a certain plant that grew nearby. He too was frightened that this clutch of birds he had already caught might be snatched away, until white faces in the window proved we weren't hungry. He showed us the deadly root, and how pretty his halfdozen unplucked pigeons were, as well as a pumpkin he wanted to sell.

Close to the rope ferry from Kerripi to Kajo Kaja, we passed several Dinkas with fishing nets and spears tall enough to fend off a crocodile or disable a hippo — which was what they were good at when not embedded in the intimacies of their cattle culture, with five hundred lowing beasts, and perhaps a lion outside the kraal to reason with; the rituals of manhood to observe; the myriad color configurations and hieroglyphic markings of each man's or boy's special display ox or bull, for him to honor, celebrate, and sing to; and the sinuous, cultivated, choreographic eloquence of its individual horns, which he tied bells and tassels to. The colors were named after the fish eagle, ibis, bustard, leopard, brindled crocodile, mongoose, monitor lizard, goshawk, baboon, elephant ivory, and so on. Like the Nuer, who were so similar, the Dinka had been famous among anthropologists but were now shattered by the war. Adventure, marriage, contentment, art, and beauty had been marked and sculpted by the visual or intuitive impact of cattle, singly and in their wise and milling, rhythmic herds, as bride wealth and the principal currency, but also the coloring registering like impressionistic altarpieces, the scaffolding of clan relations and religion. A scorched-earth policy by the Arab army and militias needed only to wipe out their cattle to disorient and dishearten the Dinka. We stopped, and people emerged from the bush. One man was burdened by a goiter the size of a bagpipe's bladder; another, by a hernia bulging like an overnight bag. They were Madi, Bari, the so-called Juba peoples, displaced by the war and siege. We had palliatives like vitamins, acetaminophen, valium, cotrimoxazole, even some iodine pills for the goiter man, although, like the hernia character, he needed surgery.

There were patients with cataracts, VD, bronchitis, scabies, nosebleeds, chest pains, Parkinson's, thrush, cellulitis, breast tumors, colon troubles, a dislocated elbow, plus the usual heartbreaking woman whose urinary tract, injured in childbirth, dripped continuously, turning her into a pariah, although it would have been as easy as the hernia for a surgeon to fix. My friend could do the elbow, with my help, and knock back an infection temporarily, but not immunize the babies or anybody because our vaccines had had no refrigeration for so long. She bestowed her smile. The line that formed, the fact that she was going to finger and eyeball everybody, was reassuring.

One night in the sunset's afterglow a man with broken eyeglasses led us to the straw church he had built, quoting Isaiah, chapter 18, and Matthew, chapter 24: "For nation will go to war against nation, kingdom against kingdom; there will be famines." Ladoku was his name, and it's no exaggeration to say that his church was constructed mainly of straw or that any big drugstore would have had ten-dollar glasses that would have helped him a lot.

On the river herons, egrets, ibises, buzzards, guinea fowl, whale-headed storks flapped every which way over the dugouts of lanky men wielding fishing tridents. They stopped at hamlets of huts only a yard above the water to let people off or leave freight, with cattle browsing in the shallows, their left horn sometimes trained a certain way by the gradual application of weights. A herder, proudly dusted with ceremonial dung-ash, was poised upon one leg, the toes of the other hooking that knee so as to jut out quite jauntily as he leaned on the point of his fighting spear, with his fishing spear dandled in his free hand. Whether Nuer or Dinka, these were people of sufficient numbers that they hadn't needed to bother learning a common language like Arabic or English to speak with other tribes, and they stared with more interest at the cargo of goodies on the deck than at the foreign or inferior strangers. The captain, although he was ethnically an Arab, had been born to shopkeepers in Malakal and had gone onto the river as a boy with his uncle, who was a pilot during the British era, and then married a daughter of the king of the Shilluk. The Shilluk were a river tribe located just northward of the Nuer and Dinka, sharing Malakal with the others as a hub. Though less numerous, they were knit rather tighter as warriors, if only because they *had* a king, so that their tough neighbors seldom messed with them.

The lions here have lost their sense of propriety. They are rattled, eating human carrion like hyenas and hauling down live individual human beings, as if there hadn't been a truce in force between the local people and the local lions for eons. Before the war, lions always knew and taught their young where they would be trespassing — what domestic beasts they shouldn't kill unless they anticipated retaliation — and people, as well, knew where it was asking for trouble for them to go. Deliberately hunting a big black-maned male might be a manhood ritual, but it was never casually undertaken with a Kalashnikov, never meaningless. Neither species was a stranger to the other or its customary habitat — whereas many of these poor refugees had been on their last legs, eating lizards, drinking from muddy puddles, wandering displaced hundreds of miles from their home ground, where they belonged. And so, on the one hand, young lionesses grew up stalking staggering people, and, on the other, soldiers in jeeps were shooting lions that they ran across with Tommy guns, for fun. No rite of passage, no conversation or negotiation was involved: no spear in the teeth, which then became a cherished necklace worn at dances.

Back in Uganda I was back in the world of AIDS. Sudan's war had kept most infected people out of the zone we'd been in, but within a few minutes I noticed that several youngsters clustered around were not healthy. They weren't wasted from starvation or fascinated by a motor vehicle, like the crowds of kids where we'd come from. I couldn't tell whether they were orphans or belonged to someone, but their stumbling, discolored emaciation meant they were dying of AIDS.

Flying over Juba we saw muzzle flashes, burning huts, and a blackish tank askew on a roadway as we banked. Inside that broken circle of machine-gun sniping and mortar explosions nobody was moving as in a normal provincial city, just scurrying for bare essentials. From the air, you could spot the positions that were crumbling and who, hunkered there, was doomed.

I ran into a couple of gaunt, drained Maryknoll nuns recuperating on a two-week Christmas holiday from their current post at Chukudum, in the Didinga Hills of borderline Sudan, where a pretty waterfall burbles down the rock bluff behind the garden of the priory, and Khartoum's Antonov bomber plane wheels over every morning looking for a target of opportunity. These nursing sisters are seasoned heroes. They're deep-dyed. You meet them on the hairiest road, coming or going from a posting, and they don't wilt. *They* don't believe that God is dead. They are wary but unflappable.

I load a truck with the standard fortified nutritional preparations and spare stethoscopes, blood-pressure cuffs, tourniquets, penlights, tongue depressors, tendon hammers, antimalarial amodiaquine, paracetamol, antibiotics such as amoxicillin, cotrimoxazole, ciprofloxacin, and doxycycline, mebendazole for worms, water purifiers, tetracycline eye ointment, ibuprofen, bandages in quantity and tape and nylon strapping, syringes and needles, scalpels, antiseptic for sterilizing, insecticide-treated mosquito netting for many people besides ourselves, IV cannulas, stitching needles and thread, umbrellas and tenting for the sun and rain, white coats for each of us to wear to give us an air of authority, and as much plastic sheeting as I have room for to shelter families in the coming rainy season. My friend Al says that children are diamonds, and he knew so from the front lines, having witnessed the successive Ethiopian and Somali famines and the Sahel droughts of the Kababish country in northern Sudan, knew that you can be nearer my God to thee without sectarianism. One Christ, many proxies.

OLIVIA JUDSON

The Selfless Gene

FROM *The Atlantic*

AT TWO A.M. ON FEBRUARY 26, 1852, the Royal Navy troop-ship *Birkenhead,* which was carrying more than six hundred people, including seven women and thirteen children, struck a rock near Danger Point, two miles off the coast of South Africa. Almost immediately, the ship began to break up. Just three lifeboats could be launched. The men were ordered to stand on deck, and they did. The women and children (along with a few sailors) were put into the lifeboats and rowed away. Only then were the men told that they could try to save themselves by swimming to shore. Most drowned or were eaten by sharks.

The heroism of the troops, standing on deck facing almost certain death while others escaped, became the stuff of legend. But the strange thing is, such heroics are not rare: humans often risk their lives for strangers — think of the firemen going into the World Trade Center — or for people they know but are not related to.

How does a propensity for self-sacrifice evolve? And what about the myriad lesser acts of daily kindness — helping a little old lady across the street, giving up a seat on the subway, returning a wallet that's been lost? Are these impulses as primal as ferocity, lust, and greed? Or are they just a thin veneer over a savage nature? Answers come from creatures as diverse as amoebas and baboons, but the story starts in the county of Kent, in southern England.

Evolving Generosity

Kent has been home to two great evolutionary biologists. In the nineteenth century, Charles Darwin lived for many years in the village of Downe. In the twentieth, William Donald Hamilton grew up catching beetles and chasing butterflies over the rolling hills near Badgers Mount.

Hamilton was a tall man with a craggy face and the tops of a couple of fingers missing from a childhood accident — he blew himself up while making explosives. He died in 2000, at age sixty-three, after an illness contracted while undertaking another risky endeavor: a trip to the Congo to collect chimpanzee feces. When I first met him, in Oxford in 1991, he had a terrific shock of white hair, rode a rickety bicycle at prodigious speed, and was preoccupied with the question of why sex is useful in evolutionary terms. (For my doctorate, I worked with him on this question.) But he began his career studying social behavior, and in the early sixties he published a trio of now-classic papers in which he offered the first rigorous explanation of how generosity can evolve and under what circumstances it is likely to emerge.

Hamilton didn't call it generosity, though; he called it altruism. And the particular behaviors he sought to explain are acts of extreme self-sacrifice, such as when a bee dies to defend the hive or when an animal spends its whole life helping others rear their children instead of having some of its own.

To see why these behaviors appear mysterious to biologists, consider how natural selection works. In every generation, some individuals leave more descendants than others. If their greater "reproductive success" is due to the particular genes they have, then natural selection has been operating.

Here's an example: suppose you're a mosquito living on the French Mediterranean coast. Tourists don't like mosquitoes, and the French authorities try to keep the tourists happy by spraying insecticide. Which means that on the coast, mosquitoes bearing a gene that confers insecticide resistance tend to leave many more descendants than those lacking it — and so today's coastal mosquitoes are far more resistant to insecticide than those that live inland.

Extreme altruists, by definition, leave no descendants: they're too busy helping others. So at first blush, a gene that promotes extreme altruism should quickly vanish from a population.

Hamilton's solution to this problem was simple and elegant. He realized that a gene promoting extreme altruism could spread if the altruist helped its close relations. The reason is that your close relations have some of the same genes you have. In humans and other mammals, full brothers and sisters have, on average, half the same genes. First cousins have, on average, an eighth of their genes in common. Among insects such as ants and bees, in which the underlying genetics work differently, full sisters (but not brothers) typically have three-quarters of their genes in common.

Hamilton derived a formula — now known as Hamilton's rule — for predicting whether the predisposition toward a given altruistic act is likely to evolve: $rB > C$. In plain language, this says that genes that promote the altruistic act will spread if the benefit (B) that the act bestows is high enough and if the genetic relationship (r) between the altruist and the beneficiary is close enough to outweigh the act's cost (C) to the altruist. Cost and benefit are both measured in nature's currency: children. "Cheap" behaviors — such as when a small bird squawks from the bushes to announce it has seen a cat or a hawk — can, and do, evolve easily, even though they often benefit nonrelatives. "Expensive" behaviors, such as working your whole life to rear someone else's children, evolve only in the context of close kin.

Since Hamilton first proposed the idea, "kin selection" has proved tremendously powerful as a way to understand cooperative and self-sacrificial behavior in a huge menagerie of animals. Look at lions. Lionesses live with their sisters, cousins, and aunts; they hunt together and help each other with child care. Bands of males, meanwhile, are typically brothers and half-brothers. Large bands are better able to keep a pride of lionesses; thus even males who never mate with a female still spread some of their genes by helping their brothers defend the pride. Or take peacocks. Males often stand in groups when they display to females. This is because females are drawn to groups of displaying males; they ogle them, then pick the guy they like best to be their mate. Again, peacocks prefer to display with their brothers rather than with males they are not related to.

Kin selection operates even in mindless creatures such as amoebas. For instance, the soil-dwelling amoeba *Dictyostelium purpureum*. When times are good, members of this species live as single cells, reproducing asexually and feasting on bacteria. But when times get tough — when there's a bacteria shortage — thousands of individuals join together into a single entity known as a slug. This glides off in search of more suitable conditions. When it finds them, the slug transforms itself into a fruiting body that looks like a tiny mushroom; some of the amoebas become the stalk, others become spores. Those in the stalk will die; only the spores will go on to form the next amoeboid generation. Sure enough, amoebas with the same genes (in other words, clones) tend to join the same slugs: they avoid mixing with genetic strangers and sacrifice themselves only for their clones.

Kin selection also accounts for some of the nastier features of human behavior, such as the tendency of stepparents to favor their own children at the expense of their stepkids. But it's not enough to explain the evolution of all aspects of social behavior, in humans or in other animals.

Living Together

Animals may begin to live together for a variety of reasons — most obviously, safety in numbers. In one of his most engaging papers, Hamilton observed that a tight flock, herd, or shoal will readily appear if every animal tries to make itself safer by moving into the middle of the group — a phenomenon he termed the "selfish herd." But protection from predators isn't the only benefit of bunching together. A bird in a flock spends more time eating and less time looking about for danger than it does when on its own. Indeed, eating well is another common reason for group living. Some predatory animals — chimpanzees, spotted hyenas, and wild dogs, for example — have evolved to hunt together.

Many social animals thus live in huge flocks or herds and not in family groups. And even if the nexus of social life is the family, the family group is itself part of a larger community. In species such as these, social behavior must extend beyond a simple "Be friendly and helpful to your family and hostile to everybody else" approach to the world. At the least, the evolution of social living requires limiting aggression so that neighbors can tolerate each other. And often the evolution of larger social groupings is accompanied by an increase in the subtlety and complexity of the ways animals get along together.

Consider baboons. Baboons are monkeys, not apes, and are thus not nearly as closely related to us as chimpanzees are. Nonetheless, baboons have evolved complex social lives. They live in troops that can number from as few as eight to as many as two hundred. Females live with their sisters, mothers, aunts, and infants; males head off to find a new troop at adolescence (around age four). Big troops typically contain several female family groups, along with some adult males. The relationships between members of a troop are varied and complex. Sometimes two or more males team up to defeat a dominant male in combat. Females often have a number of male "friends" that they associate with (friends may or may not also be sex partners). If a female is attacked or harassed, her friends will come bounding to the rescue; they will also protect her children, play with them, groom them, carry them, and sometimes share food with them. If the mother dies, they may even look after an infant in her place.

Yet friendliness and the associated small acts of affection and kindness — a bout of grooming here, a shared bite to eat there — seem like evolutionary curiosities. Small gestures like these don't affect how many children you have. Or do they?

Among social animals, one potentially important cause of premature death is murder. Infanticide can be a problem for social mammals, from baboons and chimpanzees to lions and even squirrels. During one four-year study of Belding's ground squirrels, for example, the main cause of death for juveniles was other Belding's ground squirrels; at least 8 percent of the young were murdered before being weaned. Similarly, fighting between adults — particularly in species well armed with horns, tusks, or teeth — can be lethal, and even if it is not, it may result in severe injuries, loss of status, or eviction from the group.

The possibility of death by murder creates natural selection for traits that reduce this risk. For example, any animal that can appease an aggressor or that knows when to advance and when to retreat is more likely to leave descendants than an animal that leaps wildly into any fray. Which explains why, in many social mammal species, you don't see many murders, though you do see males engaging in elaborate rituals to see who's bigger and stronger. Serious physical fights tend to break out only when both animals think they can win (that is, when they are about the same size).

Thus, among animals such as baboons, friendships mean more than a bit of mutual scratching; they play a fundamental role in an animal's ability to survive and reproduce within the group. Friendships between males can be important in overcoming a dominant male — which may in turn lead to an improvement in how attractive the animals are to females. Similarly, females that have a couple of good male friends will be more protected from bullying — and their infants less likely to be killed. Why do the males do it? Males that are friends with a particular female are more likely to become her sex partner later on, if indeed they are not already. In other words, friendship may be as primal an urge as ferocity.

Becoming Human

The lineage that became modern humans split off from the lineage that became chimpanzees around 6 million years ago. Eventually this new lineage produced the most socially versatile animal the planet has ever seen: us. How did we get to be this way?

One clue comes from chimpanzees. Chimpanzee society is the mirror image of baboon society, in that it's the females that leave home at adolescence and the males that stay where they were born. Chimpanzee communities can also be fairly large, comprising several different subcommunities and family groups. Males prefer to associate with their brothers and half-brothers on their mother's side, but they also have friendships with unrelated males. Friends hang out together and hunt together — and gang up on other males.

However, unlike baboon troops, which roam around the savanna freely intermingling, chimpanzee communities are territorial. Bands of males patrol the edges of their community's territory looking for strangers — and sometimes make deep incursions into neighboring terrain. Males on patrol move together in silence, often stopping to listen. If they run into a neighboring patrol, there may be some sort of skirmish, which may or may not be violent. But woe betide a lone animal that runs into the patrolling males. If they encounter a strange male on his own, they may well kill him. And sometimes, repeated and violent attacks by one community lead to the annihilation of another, usually smaller, one. Indeed, two of the three most-studied groups of chimpanzees have wiped out a neighboring community.

Chimpanzees have two important sources of premature death at the hands of other chimpanzees: they may be murdered by members of their own community, or they may be killed during encounters with organized bands of hostile neighbors.

Just like humans. Except that humans aren't armed with big teeth and strong limbs. Humans carry weapons and have done so for thousands of years.

On Love and War

Darwin wondered whether lethal warring between neighboring groups might have caused humans to evolve to be more helpful and kind to each other. At first, the idea seems paradoxical. But Darwin thought this could have happened if the more cohesive, unified, caring groups had been better able to triumph over their more disunited rivals. If so, the members of those cohesive yet warlike groups would have left more descendants.

For a long time, the idea languished. Why? A couple of reasons. First, it appears to depend on "group selection." This is the idea that some groups evolve characteristics that allow them to outcompete other groups, and it's long been out of favor with evolutionary biologists. In general, natural selection works much more effectively on individuals than it does on groups, unless the groups are composed of close kin. That's because group selection can be effective only when the competing groups are genetically distinct. Members of a kin group tend to be genetically similar to each other and different from members of other kin groups. In contrast, groups composed of nonkin tend to contain considerable genetic variation, and differences between such groups are generally much smaller. Moreover, contact between the groups — individuals migrating from one to another, say — will reduce any genetic differences that have started to accumulate. So unless natural selection within the groups is different — such that what it takes to survive and reproduce in one group is different from what it takes in another — migration quickly homogenizes the genetics of the whole population.

A second reason Darwin's idea has been ignored is that it seems to have a distasteful corollary. The idea implies, perhaps, that some unpleasant human characteristics — such as xenophobia or even racism — evolved in tandem with generosity and kindness. Why?

Because banding together to fight means that people must be able to tell the difference between friends (who belong in the group) and foes (who must be fought). In the mid-1970s, in a paper that speculated about how humans might have evolved, Hamilton suggested that xenophobia might be innate. He was pilloried.

But times have changed. Last year the science journal *Nature* published a paper that tested the idea of "parochial altruism" — the notion that people might prefer to help strangers from their own ethnic group over strangers from a different group; the experiment found that indeed they do. In addition, the idea that natural selection might work on groups — at least in particular and narrow circumstances — has become fashionable again. And so Darwin's idea about the evolution of human kindness as a result of war has been dusted off and scrutinized.

Sam Bowles, an economist turned evolutionary biologist who splits his time between the Santa Fe Institute, in New Mexico, and the University of Siena, in Italy, notes that during the last 90,000 years of the Pleistocene Epoch (from about 100,000 years ago until about 10,000 years ago, when agriculture emerged), the human population hardly grew. One reason for this was the extraordinary climatic volatility of the period. But another, Bowles suggests, was that our ancestors were busy killing each other in wars. Working from archaeological records and ethnographic studies, he estimates that wars between different groups could have accounted for a substantial fraction of human deaths — perhaps as much as 15 percent, on average, of those born in any given year — and as such, represented a significant source of natural selection.

Bowles shows that groups of supercooperative, altruistic humans could indeed have wiped out groups of less united folk. However, his argument works only if the cooperative groups also had practices — such as monogamy and the sharing of food with other group members — that reduced the ability of their selfish members to outreproduce their more generous members. (Monogamy helps the spread of altruism because it reduces the differences in the number of children that different people have. If, instead, one or two males monopolized all the females in the group, any genes involved in altruism would quickly disappear.) In other words, Bowles argues that a genetic predisposition for altruism would have been far more likely to evolve in groups where disparities and discord inside the group — whether over mates or food — were relatively low. Cultural differences between groups would then allow genetic differences to accumulate.

"That's Not the Way You Do It"

If Bowles's analysis is right, it suggests that individuals who could not conform, or who were disruptive, would have weakened the whole group; any group that failed to drive out such people, or kill them, would have been more likely to be overwhelmed in battle. Conversely, people who fit in — sharing the food they found, joining in hunting, helping to defend the group, and so on — would have given their group a collective advantage and thus themselves an individual evolutionary advantage.

This suggests two hypotheses. First, that one of the traits that may have evolved in humans is conformity, an ability to fit in with a group and adopt its norms and customs. Second, that enforcement of those norms and customs could have been essential for group cohesion and harmony, especially as groups got bigger (bigness is important in battles against other groups).

Let's start with conformity. This hasn't been studied much in other animals, but male baboons do appear to conform to the social regimens of the groups they join. For example, in one baboon troop in Kenya in the 1980s, all the aggressive males died of tuberculosis. The aggressives were the ones to snuff it because they'd eaten meat infected with bovine TB that had been thrown into a garbage dump; only the more aggressive males ate at the dump. After their deaths, the dynamics of the troop shifted to a more laid-back way of life. Ten years later — by which time all the original resident males had either died or moved on — the troop was still notable for its mellow attitude. The new males who had arrived had adopted the local customs.

What about humans? According to Michael Tomasello — a psychologist at the Max Planck Institute, in Leipzig, Germany, who studies the behavior of human children and of chimpanzees — children as young as three will quickly deduce and conform to rules. If an adult demonstrates a game and then a puppet comes in and plays it differently, the children will clamor to correct the puppet with shouts of "No, that's not the way you do it — you do it this way!" In other words, it's not just that they infer and obey rules; they try to enforce them, too.

Which brings me to the question of punishment.

Punishment Games

I'll be dictator. Here's how we play. An economist puts some money on the table — let's say $1,000. Since I'm dictator, I get to decide how you and I are going to split the cash; you have no say in the matter. How much do you think I'll give you?

Now, let's play the ultimatum game. We've still got $1,000 to play with, and I still get to make you an offer. But the game has a wrinkle: if you don't like the offer I make, you can refuse it. If you refuse it, we both get nothing. What do you think I'll do now?

As you've probably guessed, people tend to play the two games differently. In the dictator game, the most common offer is nothing, and the average offer is around 20 percent. In the ultimatum game, the most common offer is half the cash, while the average is around 45 percent. Offers of less than 25 percent are routinely refused — so both players go home empty-handed.

Economists scratch their heads at this. In the first place, they are surprised that some people are nice enough to share with someone they don't know, even in the dictator game, where there's nothing to lose by not sharing. Second, economists predict that people will accept any offer in the ultimatum game, no matter how low, because getting something is better than getting nothing. But that's not what happens. Instead, some people forgo getting anything themselves in order to punish someone who made an ungenerous offer. Money, it seems, is not the only currency people are dealing in.

Bring in the neuroscientists, and the other currency gets clearer. If you measure brain activity while such games are being played (and there are many variants, for the fun doesn't stop with dictators and ultimatums), you find that the reward centers of the brain — the bits that give you warm, fuzzy feelings — light up when people are cooperating. But they also light up if you punish someone who wasn't generous or watch the punishment of someone who wasn't.

Whether these responses are universal isn't clear: the genetic basis is obscure, and the number of people who've had their brain activity measured is tiny. Moreover, most economic-game playing has been done with college students; the extent to which the results hold among people from different cultures and backgrounds is relatively unknown. But the results suggest an intriguing possibility: that humans have evolved both to be good at conforming to the prevailing cultural norms and to enjoy making sure that those norms are enforced. (Perhaps this explains why schemes such as zero-tolerance policing work so well: they play into our desire to conform to the prevailing norms.)

Bringing Out the Best

If the evolutionary scenario I've outlined is even half right, then we should expect to find that there are genes involved in mediating friendly behavior. And there are. Consider Williams syndrome.

People who have Williams syndrome tend to have poor cardiovascular function and a small, pointed, "elfin" face. They are typically terrible with numbers but good with words. And they are weirdly, incautiously friendly and nice — and unafraid of strangers.

They are also missing a small segment of chromosome 7. Chromosomes are long strings of DNA. Most people have forty-six chromosomes in twenty-three pairs; you get one set of twenty-three from your mother, and the other from your father. In Williams syndrome, one copy of chromosome 7 is normal; the other is missing a small piece. The missing piece contains about twenty genes, some of which make proteins that are important in the workings of the brain. Since one chromosome is intact, the problem isn't a complete absence of the proteins that the genes encode, but an insufficiency. Somehow, this insufficiency results in people who are too nice. What's more, they can't learn not to be nice. Which is to say, someone with Williams syndrome can learn the phrase "Don't talk to strangers" but can't translate it into action.

Much about Williams syndrome remains mysterious. How the missing genes normally influence behavior is unclear; moreover, the environment has a role to play, too. But despite these complexities, Williams syndrome shows that friendliness has a genetic underpinning — that it is indeed as primal as ferocity. Indirectly, it shows something else as well. Most of us are able to apply brakes to friendly behavior, picking and choosing the people we are friendly to; those with Williams syndrome aren't. They cannot modulate their behavior. This is even odder than being too friendly. And it throws into sharp relief one of the chief features of ordinary human nature: its flexibility.

One of the most important, and least remarked upon, consequences of social living is that individual behavior must be highly flexible and tailored to circumstance: an individual who does not know whom to be aggressive toward or whom to help is unlikely to survive for long within the group. This is true for baboons and chimpanzees. It is also true for us.

Indeed, the ability to adjust our behavior to fit a given social environment is one of our main characteristics, yet it's so instinctive we don't even notice it, let alone consider it worthy of remark. But its implications are profound — and hopeful. It suggests that we can, in principle, organize society so as to bring out the best facets of our complex, evolved natures.

WALTER KIRN

The Autumn of the Multitaskers

FROM *The Atlantic*

I think your suggestion is, Can we do two things at once? Well, we're of the view that we can walk and chew gum at the same time.
— Richard Armitage, deputy secretary of state, on the wars in Afghanistan
and Iraq, June 2, 2004 (Armitage announced his resignation on
November 16, 2004)

To do two things at once is to do neither.
— Publilius Syrus, Roman slave, first century B.C.

IN THE MIDWESTERN TOWN where I grew up (a town so small that the phone line on our block was a party line well into the 1960s, meaning that we shared it with our neighbors and couldn't use it while one of them was using it, unless we wanted to quietly listen in — with their permission, naturally, and only if we were feeling awfully lonesome — while they chatted with someone else), there were two skinny brothers in their thirties who built a car that could drive into the river and become a fishing boat.

My pals and I thought the car-boat was a wonder. A thing that did one thing but also did another thing — especially the *opposite* thing, but at least an *unrelated* thing — was our idea of a great invention and a bold stride toward the future. Where we got this idea, I'll never know, but it caused us to envision a world to come teeming with crossbred, hyphenated machines. Refrigerator–TV sets. Dishwasher–air conditioners. Table saw–popcorn poppers. Camera-radios.

With that last dumb idea, we were getting close to something, as I've noted every time I've dropped or fumbled my cell phone and snapped a picture of a wall or the middle button of my shirt. Impressive. Ingenious. Yet juvenile. Arbitrary. And why a substandard camera, anyway? Why not an excellent electric razor?

Because (I told myself at the cell-phone store in the winter of 2003, as I handled a feature-laden upgrade that my new contract entitled me to purchase at a deep discount that also included a rebate) there may come a moment on a plane or in a subway station or at a mall when I and the other able-bodied males will be forced to subdue a terrorist, and my color snapshot of his trussed-up body will make the front page of *USA Today* and appear at the left shoulder of all the superstars of cable news.

While I waited for my date with citizen-journalist destiny, I took a lot of self-portraits in my Toyota and forwarded them to a girlfriend in Colorado, who reciprocated from her Jeep. Neither one of us almost died. For months. But then, one night on a snowy two-lane highway, while I was crossing Wyoming to see my girl's real face, my phone made its chirpy you-have-a-picture noise, and I glanced down in its direction while also, apparently, swerving off the pavement and sailing over a steep embankment toward a barbed-wire fence.

It was interesting to me — in retrospect, after having done some reading about the frenzied activity of the multitasking brain — how late in the process my prefrontal cortex, where our cognitive switchboards hide, changed its focus from the silly phone (*Where did it go? Did it slip between the seats? I wonder if this new photo is a nude shot or if it's another one from the topless series that seemed like such a breakthrough a month ago but that now I'm getting sick of*) to the important matter of a steel fence post sliding spearlike across my hood . . .

(*But her arms are too short to shoot a nude self-portrait with a camera phone. She'd have to do it in a mirror . . .*)

The laminated windshield glass must have been high quality; the point of the post bounced off it, leaving only a star-shaped surface crack. But I was still barreling toward sagebrush and who knew what rocks and boulders lay in wait . . .

Then the phone trilled out its normal ringtone.

Five minutes later, I'd driven out of the field and gunned it back up the embankment onto the highway and was proceeding south, heart slowing some, satellite radio tuned to a soft-rock channel called the Heart, which was playing lots of soothing Céline Dion.

"I just had an accident trying to see your picture."

"Will you get here in time to take me out to dinner?"

"I almost died."

"Well, you *sound* fine."

"Fine's not a *sound.*"

I never forgave her for that detachment. I never forgave myself for buying a camera phone.

The abiding, distinctive feature of all crashes, whether in stock prices, housing values, or hit-TV-show ratings, is that they startle but don't surprise. When the euphoria subsides, when the volatile graph lines of excitability flatten and then curve down, people realize, collectively and instantly (and not infrequently with some relief), that they've been expecting this correction. The signs were everywhere, the warnings clear, the researchers in rough agreement, and the stories down at the bar and in the office (our own stories included) revealed the same anxieties.

Which explains why the busts and reversals we deem inevitable are also the least preventable and why they startle us, if briefly, when they come — because they were inevitable for so long that they should have come already. That they haven't, we reason, can mean only one of two things. Thanks to technology or some other magic, we've entered a new age when the laws of cause and effect (as propounded by Isaac Newton and Adam Smith) have yielded to the principle of dream-and-make-it-happen (as manifested by Steve Jobs and Oprah). Either that or the thing that went up and up and up and hasn't come down, though it should have long ago, is being held aloft by our decision to forget it's up there and to carry on as though it weren't.

But on to the next inevitable contraction that everybody knows is coming, believes should have come a couple of years ago, and suspects can be postponed only if we pay no attention to the matter and stay very, very busy. I mean the end of the decade we may call the Roaring Zeros — these years of overleveraged, overextended, technology-driven, and finally unsustainable investment of our limited human energies in the dream of infinite connectivity. The overdoses, freak-outs, and collapses that converged in the late 1960s to wipe out the gains of the wide-eyed optimists who set out to "Be Here Now" but ended up making posters that read "Speed Kills" are finally coming for the wired utopians who strove to "Be

Everywhere at Once" but lost a measure of innocence, or should have, when their manic credo convinced us we could fight two wars at the same time.

The Multitasking Crash.

The Attention-Deficit Recession.

We all remember the promises. The slogans. They were all about freedom, liberation. Supposedly we were in handcuffs and wanted out of them. The key that dangled in front of us was a microchip.

"Where do you want to go today?" asked Microsoft in a mid-1990s ad campaign. The suggestion was that there were endless destinations — some geographic, some social, some intellectual — that you could reach in milliseconds by loading the right devices with the right software. It was further insinuated that where you went was purely up to you, not your spouse, your boss, your kids, or your government. Autonomy through automation.

This was the embryonic fallacy that grew up into the monster of multitasking.

Human freedom, as classically defined (to think and act and choose with minimal interference by outside powers), was not a product that firms like Microsoft could offer, but they recast it as something they *could* provide. A product for which they could raise the demand by refining its features, upping its speed, restyling its appearance, and linking it up with all the other products that promised freedom, too, but had replaced it with three inferior substitutes that they could market in its name:

Efficiency, convenience, and mobility.

For proof that these bundled minor virtues don't amount to freedom but are, instead, a formula for a period of mounting frenzy climaxing with a lapse into fatigue, consider that "Where do you want to go today?" was really manipulative advice, not an open question. "Go somewhere now," it strongly recommended, then go somewhere else tomorrow, but always go, go, go — and with our help. But did any rebel reply, "Nowhere. I like it fine right here"? Did anyone boldly ask, "What business is it of yours?" Was anyone brave enough to say, "Frankly, I want to go back to bed"?

Maybe a few of us. Not enough of us. Everyone else was going places, it seemed, and either we started going places, too — especially to those places that weren't *places* (another word they'd redefined) but were just pictures or documents or videos or boxes on screens where strangers conversed by typing — or else we'd be nowhere (a location once known as "here") doing nothing (an activity formerly labeled "living"). What a waste this would be. What a waste of our new freedom.

Our freedom to stay busy at all hours, at the task — and then the many tasks, and ultimately the multitask — of trying to be free.

> While the president continued talking on the phone (Ms. Lewinsky understood that the caller was a Member of Congress or a Senator), she performed oral sex on him.
> — The Starr Report, 1998

It isn't working, it never has worked, and though we're still pushing and driving to make it work and puzzled as to why we haven't stopped yet, which makes us think we may go on forever, the stoppage or slowdown is coming nonetheless, and when it does, we'll be startled for a moment, and then we'll acknowledge that way down deep inside ourselves (a place that we almost forgot even existed), we always knew it *couldn't* work.

The scientists know this too, and they think they know why. Through a variety of experiments, many using functional magnetic resonance imaging to measure brain activity, they've torn the mask off multitasking and revealed its true face, which is blank and pale and drawn.

Multitasking messes with the brain in several ways. At the most basic level, the mental balancing acts that it requires — the constant switching and pivoting — energize regions of the brain that specialize in visual processing and physical coordination and simultaneously appear to shortchange some of the higher areas related to memory and learning. We concentrate on the act of concentration at the expense of whatever it is that we're supposed to be concentrating *on*.

What does this mean in practice? Consider a recent experiment at UCLA, where researchers asked a group of twenty-somethings to sort index cards in two trials, once in silence and once while simultaneously listening for specific tones in a series of randomly presented sounds. The subjects' brains coped with the additional task by shifting responsibility from the hippocampus — which stores and recalls information — to the striatum, which takes care of rote, repetitive activities. Thanks to this switch, the subjects managed to sort the cards just as well with the musical distraction — but they had a much harder time remembering what, exactly, they'd been sorting once the experiment was over.

Even worse, certain studies find that multitasking boosts the level of stress-related hormones such as cortisol and adrenaline and wears down our systems through biochemical friction, prematurely aging us. In the short term, the confusion, fatigue, and chaos merely hamper our ability to focus and analyze, but in the long term, they may cause it to atrophy.

The next generation, presumably, is the hardest hit. They're the ones way out there on the cutting edge of the multitasking revolution, texting and instant-messaging each other while they download music to their iPod and update their Facebook page and complete a homework assignment and keep an eye on the episode of *The Hills* flickering on a nearby television. (A recent study from the Kaiser Family Foundation found that 53 percent of students in grades seven through twelve report consuming some other form of media while watching television; 58 percent multitask while reading; 62 percent while using the computer; and 63 percent while listening to music. "I get bored if it's not all going at once," said a seventeen-year-old quoted in the study.) They're the ones whose still-maturing brains are being shaped to process information rather than understand or even remember it.

This is the great irony of multitasking — that its overall goal, getting more done in less time, turns out to be chimerical. In reality, multitasking slows our thinking. It forces us to chop competing tasks into pieces, set them in different piles, then hunt for the pile we're interested in, pick up its pieces, review the rules for putting the pieces back together, and then attempt to do so, often quite awkwardly. (Fact, and one more reason the bubble will pop: a brain attempting to perform two tasks simultaneously will, because of all the back-and-forth stress, exhibit a substantial lag in information processing.)

Productive? Efficient? More like running up and down a beach repairing a row of sand castles as the tide comes rolling in and the rain comes pouring down. "Multitasking," a definition: "The attempt by human beings to operate like computers, often done with the assistance of computers." It begins by giving us more tasks to do, making each task harder to do, and dimming the mental powers required to do them. It finishes by making us forget exactly how on earth we did them (assuming we didn't give up, or "multiquit"), which makes them harder to do again.

Much of the problem is the metaphor. Or perhaps it's our need for metaphors in general, particularly when the subject is our minds and the comparison seems based on science. In the days of rudimentary chemistry, the mind was thought to be a beaker of swirling volatile essences. Then came classical physical mechanics, and the mind was regarded as a clocklike thing, with springs and wheels. Then it was steam-driven, maybe. A combustion chamber. Then came electricity and Freud, and it was a dynamo of polarized energies — the id charged one way, the superego the other.

Now, in the heyday of the microchip, the brain is a computer. A CPU.

Except that it's not a CPU. It's whatever that thing is that's driven to misconstrue itself — over and over, century after century — as a prototype, rendered in all-too-vulnerable tissue, of our latest marvel of technology. And before the age of modern technology, *theology*. Further back than that, it's hard to voyage, since there was a period, common sense suggests, when we didn't even know we *had* brains. Or minds. Or spirits. Humans just sort of *did* stuff. And what they did was not influenced by metaphors about what they *ought* to be *capable* of doing but very well might not be equipped for (assuming you wanted to do it in the first place), like editing a playlist to e-mail to the lover whose husband you're interviewing on the phone about the movie he made that you're discussing in the blog entry you're posting tomorrow morning and are one-quarter watching certain parts of as you eat salad and carry on the call.

Would it be possible someday — through drugs, maybe, or esoteric Buddhism, or some profound, postapocalyptic languor — to stop coming up with ideas of what we are and then laboring to live up to them?

The great spooky splendor of the brain, of course, is that no matter what we think it fundamentally resembles — even a small ethereal coliseum where angels smite demons and demons play dead, then suddenly spit fire into the angels' faces — it does a good job, a *great* job, of seeming to resemble it.

For a while.

I do like to read a book while having sex. And talk on the phone. You can get so much done.
— Jennifer Connelly, movie star, 2005

After the near-fatal consequences of my 2003 decision to buy a phone with a feature I didn't need, life went on, and rather rapidly, since multitasking eats up time in the name of saving time, rushing you through your two-year contract cycle and returning you to the company store with a suspicion that you didn't accomplish all you hoped to after your last optimistic, euphoric visit.

"Which of the ones that offer rebates don't have cameras in them?"

"The decent models all do. The best ones now have video capabilities. You can shoot little movies."

I wanted to ask, *Of what? Oncoming barbed wire?* The salesman was a believer, though — a zealot.

"Oh, yeah," he said, "as well as GPS-based, turn-by-turn navigation systems. Which are cool if you drive a lot."

"You have to look down at the screen, though."

"They're paid subscription services, you need to know, but we're giving away the first month free, and even after that, the rates are reasonable."

I shook my head. I was turning down whiz-bang features for the first time, and so had some of my friends, one of whom had sprung for a new BlackBerry that he'd holed up in his office to learn to use. He'd emerged a week later looking demoralized, muttering about getting old, although he'd just turned thirty-four.

"Those little ones there — the ones that aren't so slim, that you give away free."

"That too is an option. Mostly they're aimed at kids, though. Adolescents."

I wanted one anyway. I'd caught air in my Land Cruiser off a sheer embankment, lost my girlfriend, chucked my dream of snapping a hog-tied terrorist, and once, because of another girl — a jealous type who never trusted that I was where I said I was — I'd been forced to send on a shot of L.A. palm trees to prove that I was not in Oregon meeting up with yet another girl whom I'd drunk coffee with after a poetry reading and who must have been bombed a few weeks later when she sent me a text message at three

A.M. while I was sleeping beside the jealous girl. My bedmate heard the ring, crept out of bed, and read the message, then woke me up and demanded that I explain why it seemed to suggest we'd shared more than double espressos — an effect curiously enhanced by the note's thumb-typed dyslexic style: *Thuoght I saw thoes parkly eyes this aft, that sensaul deivlish mouth, and it took me rihgt in again, like vapmires do.*

"I'll take the fat little free one," I told the salesman.

"The thing's inert. It does nothing. It's a pet rock."

I informed him that I was old enough to have actually owned a pet rock once and that I missed it.

Here's the worst of the chilling little thoughts that have come to me during microtasking seize-ups: for every driver who has ever died while talking on a cell phone (researchers at the Harvard Center for Risk Analysis estimate that some 2,600 deaths and 330,000 injuries may be caused by drivers on cell phones each year), there was someone on the other end who, chances are, was too distracted to notice. Too busy cooking, NordicTracking, fluffing up his online dating profile, or — most hauntingly of all, I'd think, for a listener destined to discover that the acoustic chaos he'd interpreted as the other phone going out of range or perhaps as a network-wide disturbance triggered by a solar flare, was actually a death, a human death, a death he had some role in — sitting on the toilet.

Trading securities.

Or would watching streaming pornography be worse?

Not that both of these activities can't be performed on the same computer screen. And often are — you can bet on it. In bathrooms. Even *airport* bathrooms, on occasion. In some of which, via radio, the latest business headlines can be monitored, permitting (in theory and therefore in *fact,* because, as the First Law of Multitasking dictates, any two or eight or sixteen processes that *can* overlap *must* overlap) the squatting day trader viewing the dirty Webcast (while on the phone with someone, don't forget) to learn that the company he just bought stock in has entered merger talks with *his own employer* and surged almost 20 percent in under three minutes!

"Guess how much richer I've gotten while we've been yakking?"

he says into his cell, breaking his own rule about pretending that when he's on the phone, he's on the phone. Exclusively. Fully. With his entire being.

No reply.

Must be driving through a tunnel.

I've been fired, I've been insulted in front of coworkers, but the time I flew thousands of miles to meet a boss who spent our first and only hour together politely nodding at my proposals while thumbing out messages on a new device, whose existence neither of us acknowledged and whose screen he kept tilted so I couldn't see it, still feels, five years later, like the low point of my career.

> This is the perfect "one plus one equals three" opportunity.
> — Robert Pittman, president and COO of America Online, on the merger between AOL and Time Warner, 2000

There may be a financial cost to multitasking as well. The sum is extremely large and hard to vouch for, the esoteric algorithm that yielded it a puzzle to all but its creator, possibly, but it's one of those figures that's fun to quote in bars.

Six hundred and fifty billion dollars. That's what we might call our National Attention Deficit, according to Jonathan B. Spira, who's the chief analyst at a business-research firm called Basex and has estimated the per annum cost to the economy of multitasking-induced disruptions. (He obtained the figure by surveying office workers across the country, who reported that some 28 percent of their time was wasted dealing with multitasking-related transitions and interruptions.)

That $650 billion reflects just one year's loss. This means that the total debt is vastly higher, since personal digital assistants (the devices that, in my opinion, turned multitasking from a habit into a pathology, which the advent of Bluetooth then rendered fatal and the spread of wireless broadband made communicable) are several annums old. This puts our shortfall somewhere in the trillions — even before we add in the many billions that vanished when Time Warner and AOL joined their respective corporate missions, so ably accomplished when the firms were separate, into one colossal mission impossible.

And don't forget to add Enron to the tab, a company that seemed to master so many enterprises, trading everything from energy to weather futures, that the Wall Street analysts' brains froze up trying to "recontext" (another science term) what looked at first like a capitalist dynamo as the street-corner con that it turned out to be. Reports suggest that the illusion depended nearly as much on cunning set design as it did on phony accounting. The towering stack of Broadway stages that Enron called its headquarters — with its profusion of workstations, trading boards, copiers, speakerphones, fax machines, and shredders — made visiting banker-broker types go snow-blind. When the fraud was exposed, the press accused the moneymen of overestimating Enron. In truth, they'd underestimated Enron, whose hectic multitasking front concealed the managers' Zenlike focus on one proficiency, and only one.

Hypnotism.

Which is easy to practice on an audience whose brains are already half dormant from the stress of scheduling flights on fractionally owned jets and changing the tilt and speed of treadmills according to the shifting readouts of miniature biofeedback monitors strapped around their upper arms.

What has the madness of multitasking cost us? The better question might be: what hasn't it?

And the IOUs keep coming, signed at the bottom with millions of our names. We issued this currency. We're the Federal Reserve of the attention economy, the central bank of overcommitment, keeping the system liquid with adrenaline. The problem is that we, the bankers, are also the borrowers. That's multitasking for you. It moves in circles. Circles that we run around ourselves, as we try to pay off the debts we owe ourselves with funny money engraved with our own faces.

Here's one item from my ledger:

Cost of pitying Kevin Federline while organizing business trip online and attaching computer peripheral: $279.

Federline — I know. A mayfly on the multimedia river who, now that he has mated, deserves to break back up into pixels. That he hasn't means pixels are far too cheap and plentiful, particularly on the AOL welcome page, where for several months last year Federline's image was regularly positioned beside the icon I click to get my e-mail. With practice, I learned to sweep past him the way the queen sweeps past her guards, but one afternoon his picture triggered a brainslide that buried half my day.

What the avalanche overwhelmed was a mental function that David E. Meyer, a psychology professor at the University of Michigan, calls "adaptive executive control." Thanks to Federline, I lost my ability, as Meyer would say, to "schedule task processes appropriately" and to "obey instructions about their relative priorities."

Meyer, it's worth noting, is a relative optimist among the researchers studying multitasking, since he's convinced that with enough practice, some people can learn to perform two tasks simultaneously as successfully as if they were doing them sequentially. But "enough practice" turns out to mean at least 2,000 tries, and I had just the one chance at the cheap fare to San Francisco that I'd turned on my laptop to reserve, only to be distracted by the picture of Federline winking at me from one browser window over.

The photo, a link explained, was taken while Federline was taping a TV show and happened to peer down at his phone, only to learn that what's-her-hair, his wife, the psycho, bad-mother rehab-escapee (I had last caught up on her misadventures weeks or months before while waiting out an eBay auction for an auxiliary hard drive "still in box"), had sent him a text message asking for a divorce. Federline's face looked as raw as a freshly unbandaged plastic-surgery patient's, but the aspect of the photo that grabbed me (as the promotional fare hovered in the ether, still unbooked and up for grabs) was the idea I suddenly entertained about its origins. The picture of Federline in cell-phone shock had been snapped on the sly by another phone, I sensed, and possibly by a hanger-on whom Federline regarded as a "bro." It also seemed plausible that after the taping, Federline bought dinner for this Judas — who, in my reconstruction of events, had already beamed the spy shot to a tabloid and been wired big money in return. If so, he was probably richer than Federline, who depended for funds on the wife who'd just dumped him.

This thought sequence caused me to remember the hard drive — still sitting unopened in a closet — that I'd bought in that Internet auction way back when, while catching up on the Hollywood gossip news. Here's the mental flow chart: Federline dumped > story about his prenuptial with Britney Spears > story read during eBay auction > time to get some use out of my purchase.

Removing the hard drive from its shell of molded Styrofoam

sloppily wrapped in masking tape stirred serious doubts about the seller's claim that the gadget was unused. This put me in a quandary. Should I send the hard drive back? Blackball its seller on a message board? Best to test it first. I riffled through drawers to find the proper cable, plugged the device into a USB port, and only then became aware of the fluorescent Post-it note stuck in the corner of my laptop screen. "Grab discount SF fare," the note read. Where had it gone? Where had *I* gone, rather? How could a piece of paper in a color specially formulated to signal the brain *Important! Don't Ignore!* be upstaged by a picture of a sad minor celebrity? If the Post-it note had been a road sign warning of a hairpin mountain curve and Federline's photo a radio interview, I and my car would be rolling down a cliff now.

Back to the San Francisco ticket, then. I brought up the main Expedia/Orbitz/Travelocity page and typed in the code for the San Francisco airport, which I couldn't believe I got wrong. To fix it, I was forced to use one of those drop-down alphabetized lists that the highlight line always moves too fast through, meaning I click my mouse several entries too late. Seattle this time. I scrolled back up.

All tickets sold out.

The scientists call this ruinous mental lurching "dual task interference," or just plain bottlenecking. I call it the reason Kevin Federline cost me a cheap flight to San Francisco. (It also explains, perhaps, why sexual threesomes are often disappointing.)

I just wish the military understood the concept. They might understand then why "walking and chewing gum" in Afghanistan and Iraq is no way to catch bin Laden.

My hunch is that when we look back on it someday, at our juggling of electronic lives and the array of subtly different personas that each one encourages (we're terse when texting, freewheeling on the phone, and in some middle state while e-mailing), the spectacle will appear as quaint and stylized as those scenes in old movies of stiff-backed lady operators, hair in bobby pins, rapidly swapping phone jacks from hole to hole as they connect Chicago to Miami, reporter to city desk, businessman to mistress. Such scenes were, for a time, cinematic shorthand for the frenzy of modern life, but then communications technology changed, and those operators lost their jobs.

To us.

We've got to be patient and committed [in Iraq], but we've got to multitask
. . . We've got to talk about Iran — Iran is more dangerous than Iraq — and
we have got to get the job done in Afghanistan and in Pakistan.
        — Rudolph Giuliani, Republican presidential candidate, July 2007

The night the bubble finally popped for me began when I pushed a
button on my hospital bed to summon the gray-haired night nurse.
To convey my appreciation when she arrived and to help establish a
relationship that I hoped would lead her to agree with me that my
morphine drip was far too slow, I did as the gurus of management
urge executives to do when they engage in important negotiations.
I "reallocated" my "presence" and "enriched" my "medium." I removed my headphones, closed my book, aimed the remote and
clicked off the TV, and looked the old woman in the eye.

"What?" she said.

Her question came too quickly. Because of the way the human
brain works — always lagging slightly, always falling a bit behind itself when it has to drop many things, one thing at a time, and refo-
cus on a new thing — my attention had not yet caught up with my
expression. Also, perhaps because of the way that morphine works,
I was unnaturally aware of the mechanisms inside my mind. I could
actually feel the neurological switching, the mental grinding of
fine, tiny gears that makes multitasking such an inefficient, slow,
error-prone, tiring way to get things done.

"Still hurts," I finally said. "Wondering if you'd shorten up the intervals." I left out the *I*'s, text message–style, because that's how
people in agony communicate. Teenagers, too — but aren't they
also in agony, with the shy self-consciousness of partials who don't
show all their cards, out of fear that they haven't yet drawn many
worth playing?

The nurse made a face that the gurus would call "equivocal"
— meaning that it can support conflicting interpretations, even
in a real-time, face-to-face, "presence-rich" exchange — and then
glanced down at the iPod on my blanket.

"Music lover?"

"Book on tape," I said.

"You can do those both at once?" She eyed the real book lying on
my lap.

"Same one," I said. "I like to double up."

"Why?"

I had no answer. I had a comeback — *Because I can, because it's possible* — but a comeback is just a way to keep things rolling when perhaps they ought to stop. When the nurse looked away and punched in new instructions on the keypad attached to my IV stand, I heard her thinking, *No wonder this guy has kidney stones. No wonder he's so hungry for narcotics.* She turned around in time to see my hands moving from the book they'd just reopened to the tangled wires of the headphones.

"I'm grateful that you came so quickly and showed such understanding," I said, not textishly, relaxing my syntax to suit the expectations of the elderly.

"Maybe more dope will be just the thing," the nurse said, shedding equivocation with every word, as a dreamy warmth spread through my limbs and she soft-stepped out and shut the door. When I woke in the wee hours, my book, in both its forms, had slid off the bed onto the floor, the TV remote was lost among the blankets, and the blinking "sleep" indicator of the laptop computer I've failed to mention (delivered to my bedside by a friend who'd shared my delusion that even twenty-five-bed Montana hospitals must offer wireless Internet these days) was exhaling onto the walls a lovely blue light that tempted me never to boot it up again.

That night last May, as I drowsed and passed my stones, the mania left me, and it hasn't returned.

What happened to the skinny brothers' car-boat was that it sank the third time they took it fishing. It cracked down the length of its hull, took on water, then nose-dived into the sandy bottom, leaving its revved-up rear propeller sticking up two feet out of the river, furiously churning air until its creators returned in a canoe and whacked it silent with a crowbar.

The catastrophe, visible from half the town, was the talk of the party line that night, with most of the grownups joining in one pooled call that was still humming when I was sent to bed.

"Where do you want to go today?" Microsoft asked us.

Now that I no longer confuse freedom with speed, convenience, and mobility, my answer would be: "Away. Just away. Someplace where I can think."

ANDREW LAWLER

First Churches of the
Jesus Cult

FROM *Archaeology*

AS DUSK APPROACHES, Korean pilgrims in white baseball caps blow horns and sing hymns atop Tel Megiddo. This crossroads in northern Israel — also known as Armageddon — is where the New Testament says the final battle pitting good against evil will begin. Below the huge mound, tour buses idle, throngs of visitors buy postcards, and a nearby McDonald's does a thriving business at its drive-through window.

On the opposite side of the busy highway are the grim brick walls and coiled barbed wire of a high-security prison. It is an awkward place for an important archaeological site. In contrast to the mound, visitors are not welcome here. Even archaeologists must apply well in advance for access — something I wasn't granted — so I am left standing outside the gates with Yotam Tepper of the Israel Antiquities Authority. The mosaic floor that he and a team of inmates discovered under the prison yard may mark one of the earliest known places of Christian worship.

Although the site may date to a full century before the Roman emperor Constantine issued the Edict of Milan, transforming Christianity from a disparate group of Jesus-worshiping cults to a powerful state religion in A.D. 313, those early followers of the controversial faith weren't hiding their beliefs. "There were Samaritans and Jews and Romans and Christians all living together in just this small place," says Tepper. A Roman soldier paid for the mosaics, and members of the congregation may even have baked bread for Rome's sixth legion, stationed nearby.

The find at Megiddo is a key piece of evidence in a radical re-thinking of how Christianity evolved during its first three centuries, before it was backed by the might of empire. Until recently, schol-ars had to rely on ancient texts that emphasize the vicious persecu-tion of the church — think lions dining on martyrs in Rome's Col-osseum. A growing body of archaeological data, however, paints a more diverse and surprising picture, in which Christians thrived alongside Jews and the Roman military. These finds make this "a definitive time in our field" since they appear to contradict the lit-erary sources on which historians have long depended, says Eric Meyers, a biblical archaeologist at Duke University.

Megiddo is only the latest in a series of recent digs in the Near East revealing a more complex history of the early Christian era. Near the Red Sea in the Jordanian city of Aqaba, archaeologists have uncovered what the dig director, Thomas Parker of North Carolina State University, argues is a pre-Constantinian prayer hall. At Capernaum, just an hour's drive from Megiddo, Franciscan monks believe they have excavated a pilgrimage site dating to as early as the first century A.D. on the shores of the Sea of Galilee. Such discoveries are unusual; the only undisputed early Christian worship site is at Dura Europas, on the Euphrates River in modern Syria, which was excavated in the 1920s and '30s by French and American teams. How the most recently discovered sites were used and dated, however, is hotly contested.

Formal churches were rare before A.D. 325, when Constantine convened the Council of Nicea, formalizing many church prac-tices, and embarked on a building campaign that used the Roman basilica — a spacious rectangular enclosed space, typically with an apse and an altar on one end — as the model for Christian places of worship. The basilica became the standard still used for churches around the world.

Before that innovation, however, Christians would gather in a *domus ecclesia,* or house church. Eager to keep a low profile dur-ing uncertain times, many Christian communities met in homes throughout the first centuries to celebrate rituals such as the Eu-charist, which used wine and bread to recall Christ's sacrifice and to bind the community of believers together. In a letter to the Romans, Saint Paul mentions "the church that is in their house," and numerous other early writers cite homes where congregations met. "This type of architecture was quite private, so it was not visibly a Christian building," says Joan Taylor, a historian at University College London. "Otherwise, it might get smashed and you might get killed."

That was a legitimate fear. The Jewish high council, according to the New Testament, ordered the death of the first Christian martyr, Stephen. Christians — who still were seen as a Jewish sect — refused to join Jews in the Bar Kokhba revolt against the Romans in A.D. 132–135. Judged as traitors by the Jewish community, they were killed in retribution. After the revolt, however, the decimated Jewish population posed far less of a threat than the Romans. Nero had already scapegoated Christians for burning Rome in A.D. 64; Emperor Decius (A.D. 249–251) had pursued lay Christians as well as clergy; and Diocletian and Galerius had infamously persecuted Christians at the end of the third and the beginning of the fourth century A.D. There is little doubt that Christians suffered terribly during the religion's early days. But the evidence from Near Eastern digs, combined with new thinking about the Roman Empire, demonstrates that there were substantial periods when Christians were tolerated, accepted, and even embraced by their tormentors.

This was indisputably the case at Dura Europas, a formidable city and Roman garrison that guarded the eastern frontier of the empire. Excavations in the 1930s revealed a domus ecclesia that includes an inscription dating it to A.D. 231 — the only Christian house church that scholars agree predates Constantine. The house church was located near the city gate, where Roman soldiers would have been stationed. "There's no way the Romans didn't know about the Christians," says Simon James, an archaeologist at the University of Leicester.

For decades the house church has remained an archaeological oddity. New clues, however, have been emerging far to the south, at Capernaum along the Sea of Galilee in Israel, where Franciscan scholars have been excavating a site for the past century. They believe it was the house of Peter and other apostles; Jesus is said to have lived here and taught at the local synagogue. Today a squat and ugly modern concrete church hovers above the house. Visiting Italian nuns and Nigerian pilgrims peer down through the church's glass floor at the foundations of the octagonal shrine built a century or so after Constantine legalized Christianity. The octagon was a typical shape for shrines and places of importance, from Roman tombs to the Dome of the Rock. Below the Capernaum structure, the excavators found eleven floors, layered one on top of the other, dating from the second century B.C. through the fourth century A.D., says Michele Piccirillo, a Franciscan archaeologist.

Piccirillo's office is a high-ceilinged room just off the Via Dolorosa in Jerusalem, with a bare bulb illuminating religious paintings and stacks of books. He makes strong and bitter coffee as he lays out the case for Capernaum as one of Christianity's most ancient places of worship. Digging through his papers, he points out the evolution of the house. He notes that the early layers include lamps and cooking pots, while from the second century A.D. on, only lamps have been found — circumstantial evidence that the site may have been transformed from a private home into a place of pilgrimage or worship. And some bits of plaster in the central room show graffiti by Christians, including the name Peter and references to "Christ" and "Lord" in Aramaic, Greek, Latin, and Syriac. "There is continuity — this house eventually was used as a church," he says. He believes the domus ecclesia dates from at least the third century.

Other archaeologists disagree with this interpretation. Joan Taylor, who closely examined the data, believes the site was not used for worship until the fourth century. But Eric Meyers is impressed with the evidence. "There is no doubt that the graffiti suggest early Christian pilgrims venerated the site," he says. "The excavators have been very, very responsible — they're not making this up." But he adds that Franciscans like Piccirillo "have a vested interest in proving the antiquity of holy sites." What is not in dispute, however, is the existence of an elaborate synagogue across the street, dating to the same time as the octagonal building. The Franciscans believe it was built on the foundation of an earlier Jewish house of worship dating to the first century A.D. — and possibly the same one in which Jesus is said to have preached. Whether or not the monks have found Peter's house, it is clear that Jews and Christians coexisted peacefully here.

Farther to the south, in Jordan, the team led by Thomas Parker uncovered another candidate for a pre-Constantinian church in the late 1990s. Located just a short walk from the Red Sea in the port of Aqaba, the small site today is surrounded by busy streets and hotels in this popular seaside resort. Like Dura Europas, the city in Roman times was a thriving center of trade at the edge of the empire — and an important military post. Unlike a scattering of other archaeological sites in this city, there are no signs yet explaining the potential significance of the mud-brick structure that lies crumbling in the sun, protected by a short wire fence. More than one hundred coins, the latest dating to the last decade in the reign of Constantinius II (A.D. 337–361), were found in the building, which measures 85 by 53 feet. Based on the coins and pottery, Parker estimates that the building was constructed in the late third or early fourth century A.D. — though he says a post-325 date is not out of the question.

Given the east-west orientation, basilica-like plan, glass oil-lamp fragments, and a cross found in a grave in a nearby cemetery, he argues that the building was a formal church rather than a domus ecclesia. The theory has yet to win many supporters, but scholars are eager to see his final publication of the find, which should be out this year. "I am skeptical," says Jodi Magness, an archaeologist at the University of North Carolina at Chapel Hill, who specializes in the period and is digging just across the border in Israel. "I haven't seen anything yet that persuades me."

Magness has her own potential candidate for a pre-Constantinian church in southern Israel at a site called Yotvata, a Roman fort that was built around A.D. 300. In 2006 her team found a semicircular niche cut into the fort's wall flanked by two pilasters and an inscription that may be a Christian prayer. The niche was likely built in the early fourth century, but a more precise date will require further excavation — including the removal of a British police station that was built over it in the 1930s.

The controversy surrounding the church at Megiddo began in 2003, when prisoners were assigned to expand the buildings housing Christian and Muslim Palestinian prisoners. When the crew working in the interior yard hit archaeological remains, prison officials alerted the Israel Antiquities Authority, which put Tepper, a graduate student at Tel Aviv University, in charge of the salvage effort. He conducted the work primarily with a team made up of seventy prisoners.

Like Dura Europas and Aqaba, Megiddo was full of Roman soldiers. And like Capernaum, it was primarily a Jewish town. It is situated on a strategic spot between the Mediterranean coast and the Sea of Galilee, its bloody future as the site of the last battle between good and evil forecast by the New Testament book of Revelation reflects its past: here battles raged involving Egyptians, Canaanites, Assyrians, Greeks, Romans, Turks, and British. But during Roman times, it was the site of a Jewish village called Kefar 'Othnay, a Roman legion camp, and eventually a Byzantine city called Maximianopolis.

"It was a small village with nothing special," says Tepper. The settlement, likely founded by Jews or Samaritans in the second half of the first century, covered about fifteen acres and was located next to a Roman legion base. In late 2005, as he was wrapping up the dig, Tepper came across the remains of a building on the edge of the village closest to the Roman camp. The building had four wings, an exterior courtyard with bread ovens, and a series of rooms opening onto an interior courtyard. In the western wing Tepper's team uncovered a hall measuring 5 by 10 yards and oriented north to south. In the middle of the hall, they found four mosaic panels with inscriptions surrounding a podium. Two panels are decorated with simple geometric patterns; a third is slightly larger, with Greek inscriptions on each end. The fourth shows two flopping fish — a tuna and a sea bass — circled by squares, triangles, and diamonds with a large inscription on one end.

Tepper faxed images of the mosaics to Leah di Segni, an epigrapher at Hebrew University who was working from her third-floor walk-up apartment in West Jerusalem. At first, she says, she assumed the mosaics were part of a temple to Mithras, a Persian god popular with Roman troops from the empire's eastern frontier to Scotland. Di Segni translated one inscription as "Gaianus, also called Porphyrius, centurion, our brother, has made the pavement at his own expense as an act of liberality. Brutius has carried out the work." A second inscription is a memorial to four women with common Greek names. But the third inscription was the stunner: "The god-loving Akeptous has offered the table to God Jesus Christ as a memorial."

She immediately phoned Tepper and told him to look for Roman pottery. He promptly found shards and coins that he says date the site to the early third century. They found more than one hundred coins in the complex, one-third of which date to the second and third centuries A.D. and the remaining two-thirds to the fourth century. Almost all of the early coins come from the hall, including several in pristine condition from the reigns of the emperors Elagabalus (A.D. 218–222) and Severus Alexander (A.D. 222–235). "These coins," Tepper says, "should probably be associated with the founding of the building." The latest one, he notes, is dated to Diocletian's reign in the late third century. The absence of any post-Diocletian coins may mean that the building was abandoned in the fourth century, says Tepper. He also says he has Roman pottery that confirms his conclusion.

Most of the jar fragments in the complex appear to be from the third century A.D., with the latest dating to the early fourth century. Pottery fragments found alongside and below the mosaic floor are no later than the third century, he adds. Two stone stamps that were used by the bakers of the Roman legions to mark the bread they made were found in the complex, another sign that soldiers may have been Christians at a time when the faith was officially outlawed.

Tepper's conclusions have been greeted skeptically by senior archaeologists, such as Magness and Piccirillo. "There are a lot of early coins — so what?" says Magness, who notes that the area under the mosaic floor, which might yield critical dating material, has yet to be excavated. "I don't think they have convincing evidence," she adds. Piccirillo agrees. An expert in Byzantine mosaics, he believes their style indicates they could be as late as the fifth century.

Others are more intrigued. "I'm open to Megiddo as a third-century site," says Taylor. "It's idiosyncratic," she adds, since it does not fit the model of Christian churches during and after the time of Constantine. Those structures are easily recognizable by their basilica shape, with an altar at the east end and the main entrance to the west. "This is a time before all the dictates come from above," says Taylor. And Meyers, a pottery expert, says that while everyone is awaiting a final publication, he is convinced that the shards are distinctively mid-Roman rather than from a later era.

If Megiddo does prove to be an early prayer hall, then it will lend strength to the growing view among scholars that the early church in the Holy Land was highly diverse during the two centuries between the death of Jesus and Constantine's edict. "The traditional view was that early Christianity was not licensed, that it had to hide," says Taylor. "That's shifting to a recognition that there were periods of persecution followed by periods of peace." And those well-documented periods of persecution might have had spotty results. Decrees issued from Rome, Taylor says, might have little impact at the fringes of a vast empire, at places like Megiddo and Dura Europas.

Meyers agrees. He also believes the Megiddo site is evidence that scholars need to rethink the idea that the Holy Land was largely devoid of Christians after the Roman destruction of Jerusalem and subsequent Jewish revolts. According to Meyers, the archaeological evidence points to a complex and closer relationship between early Christians and Jews. Despite Byzantine decrees persecuting Jews, he notes that impressive synagogues sprang up around the empire at places like Capernaum. "The two sister religions have an often robust and positive" relationship, says Meyers. He believes the excavations show that it goes back to Christianity's early days.

How Roman soldiers influenced the evolution of early Christianity remains an open question. Though the Roman army was often the weapon used to smash Christian places of worship, soldiers were also drawn to a host of eastern cults, such as Mithraism and Christianity. "A lot of soldiers regarded it as sensible to get on the right side of the local deities," says James.

Meanwhile work at the Megiddo site has stopped. Israeli officials would like to move the prison, but there is no budget to do so. There are not even funds to finish the excavations and conserve the site. The idea of turning the area into a major tourist destination — the nearby Tel Megiddo already draws hundreds of pilgrims each day — appears to be on indefinite hold. Standing outside the prison gate, Tepper says that the money and jobs involved make moving the prison difficult. He is currently busy with other salvage excavations around the Sea of Galilee. By now the sun is setting and the tourist buses have all left Tel Megiddo. Tepper gives the prison walls one last glance and climbs into his battered Jeep as the gate opens briefly — but only to let in a new batch of prisoners.

JON MOOALLEM

A Curious Attraction

FROM *Harper's Magazine*

Spooky Action at a Distance

EVEN SIR ISAAC NEWTON recognized that the idea of gravity —
that one object can instantaneously yank another through total
nothingness and from extremely far away — was so counterintuitive, "so great an absurdity," that no intelligent person could be expected to stand for it. It lacked the elegant horse sense of the old
ideas, such as the rock falling to Earth because it is made of earth.
Newton was able to imagine this black magic moving the apple because, as one biographer admiringly writes, "he embraced invisible
forces." And he did so more promiscuously than we choose to remember. The inventor of modern gravity was also a fanatical alche-
mist. It's just that in the case of universal gravitation, the invisible
force he embraced turned out to be real.

"Most of the fundamental ideas of science are essentially simple,
and may, as a rule, be expressed in a language comprehensible
to everyone," Albert Einstein once wrote. Yet in 1915, Einstein's
theory of general relativity pulled gravity even further from the
realm of common sense, entangling it in recondite mathematics
and dreamlike geometry.[1] Initially, it didn't seem to matter whether
Einstein was right; his masterpiece was attacked for not being *popu-
list* enough.

"A great and serious retrograde step," one Princetonian called it.
"All previous physical theories have been thus intelligible . . . [to]
the whole race of man." A flummoxed Columbia professor announced that general relativity smacked of Bolshevism and must be
stopped.

In 1919 astronomers made the first significant observation to support general relativity when they watched starlight bending through space-time curved by the sun. "It is not possible to put Einstein's theory into really intelligible words," the *New York Times* noted in its coverage of the event. One scientist tried, but it was as though he'd had a conversion experience into a very small cult. What this new observation showed, he told the *Times,* was that parallel lines could eventually meet, that three angles of a triangle needn't add up to 180 degrees, and that "a circle is not really circular." The editors headlined the story, in part: STARS NOT WHERE THEY SEEMED OR WERE CALCULATED TO BE, BUT NOBODY NEED WORRY.

As in Newton's time, a seemingly unbridgeable rift opened between everyman's visceral experience of gravity and the specialist's explanation of it, and the laymen didn't always accept this estrangement with the wry dispassion of the *Times.* Somehow it just didn't seem right that this ubiquitous force affecting us all should be understood by only a select few. A man named George Francis Gillette published *Orthod Oxen of Science,* claiming that Einstein's "self created fairyland of 4th dimensional space" was "the funniest mental bellyache inflicted on science." It was also "utter tommy-rot." Gillette offered a number of ideas in place of Einstein's hooey. Among them was "The All Cosmos Doughnut."[2]

Despite the initial resistance, however, physics rapidly fell in love with general relativity. More observations supported it, and its predictions seemed to operate on too astronomical a scale to be checked directly with any technologically feasible experiments. Einstein's gravity came to stand as a kind of untestable truth.

Nevertheless, as physics has worked to refine its understanding of gravity, a sanguine subculture of amateurs and outsiders has continued to needle the accepted thinking for weaknesses, setting out not to better define this invisible force but to overcome it. Much of this recreational research asks whether we might finally be able to do something with this power and not simply suffer from it. Today the Web site American Antigravity serves as a clearinghouse for this quasi-scientific uprising, and the mere copiousness of its holdings has a way of making the incredible feel almost inevitable. Surely a "cure" for gravity will arise from something here: the "Angelina VI" ball-lightning generator, the Beifeld-Brown Effect, the internal plasma expansion/contraction engine, the Impulse Gravity

Generator based on a charged $YBa2Cu3O7-y$ superconductor with composite-crystal structure, or the Quantum Vacuum Pathway Theory, which "describes a plausible mechanism" for "spooky action at a distance."[3]

We are suspenders of disbelief, easily enchanted by possibility, addicted to wonder. So whatever measure of faith we harbor in the fallibility of gravity may, like our faith in so many things, be sustained not by facts or lack of facts as much as by the sheer strength of our longing for it to be so. After all, what clearer vision of joyousness and freedom is there than a band of jumpsuited astronauts in zero gravity tumbling upward and over themselves like giddy pinwheels? "Of all the natural forces," wrote the futurist Arthur C. Clarke, "gravity is the most mysterious and the most implacable." Yet Clarke also assumed that the high Himalayan peaks would one day be as saturated with tourists as the beaches of Cannes, once personal gravity control was perfected. "The Sherpas and Alpine guides will, of course, be indignant," he wrote. "But progress is inexorable."

After a century of science fiction on the subject, a life unfettered by gravity may feel strangely like an entitlement. Indeed, there is mounting suspicion even in the most respectable spheres of physics that the force we call "law" — this fact evidenced by every dropped paper clip and sunset, as frank as a falling anvil — may not actually be settled at all.

Gravity Radio

In 2002, at a Princeton physics symposium titled "Science and Ultimate Reality," Raymond Chiao announced his plans to turn gravity into both electricity and light, and electricity and light into gravity. For more than twenty years, Chiao had been laboring over a revolutionary communications apparatus that would work by linking these disparate forces. He'd originally called the device a "gravitational radiation antenna," but his son-in-law eventually came up with a better name: gravity radio.

Chiao is sixty-six, an MIT alumnus, and a fellow of both the Optical Society of America and the American Physical Society. At the time of the Princeton symposium, he was still a celebrated professor of physics at the University of California, Berkeley. As Chiao spoke, he described his contraption in the drab and equation-encumbered language of physics. Yet the rattling wondrousness of gravity radio and the world into which it could deliver us was impossible to miss. What could gravity radio do? Gravity radio could beam unstoppable, information-rich gravity waves straight through the soil, crust, and core of Earth to be received — unattenuated and unscathed — by gravity radios on the opposite side of the planet. Relaying a flimsy signal between satellites might be a serviceable way to dial China, but here was a phone call of unwavering directness. Point gravity radio up, and gravity radio could send e-mails to Venus or farther, at the speed of light. Gravity radio could broadcast *Fresh Air* to Alpha Centauri. Gravity radio could pick up gravity waves still emanating from the big bang — which is to say, gravity radio could be a baby monitor tuned to the infancy of time.

"It is hard to know whether the assembly was more astonished by the idea that this might be possible," one observer in attendance noted, "or by Chiao's lack of concern for his own reputation." No one has ever even seen a gravity wave. Einstein predicted that these ripples in space-time are sent coursing through the cosmos any time an object moves. Yet Einstein himself had trouble believing this at first; then, conveniently, he decided that if gravity waves did exist (we now know they almost certainly do), they would be such slight disturbances that we'd never be able to detect them. For half a century, in fact, no one bothered to try.

As it happens, while Chiao was delivering his paper, science's best hope of pinning down gravity waves was being readied to switch on after thirty years of development. The experiment is called LIGO, the Laser Interferometer Gravitational-Wave Observatory, and it involves multimillion-dollar arrays of lasers streaking through sets of 2.5-mile-long stainless-steel tubes in Washington State and the Louisiana woods. Still, even with these massive antennae, LIGO might barely be able to detect the universe's most robust gravity waves, such as those emitted from violent supernovae. Gravity radio would practically fit on a tabletop.

"This is the too-good-to-be-true argument," a colleague of Chiao's told me. "It threatens everyone else's intuition." When a reporter for the magazine *New Scientist* visited Chiao at Berkeley in 2003, he found gravity radio on the floor. It was a slab of superconducting ceramic, a couple of empty paint cans for insulation, some polystyrene cups, and bits of wood, all disassembled and piled into a cardboard box. It looked like trash. Chiao was funding gravity radio himself, ordering materials from high school suppliers. A peer-reviewed journal, he said, had rejected one of his gravity-radio papers without explanation.

When I called Chiao two years later, he seemed understandably ambivalent about explaining the promise and controversies of gravity radio to a novice. He still hadn't performed any experiments; he was refining his theories. "Let me put it this way," he said. "It's a long shot. But I think it's worth continued research."

Several weeks later, seeming more confident about gravity radio than ever, Chiao suddenly left his job at Berkeley to build it. I was welcome to visit his new lab, he said, if I was willing to make the drive.

Our Enemy Number One

I'd first learned about gravity radio while tracing the loose ends of a figure whose life I was coming to see as both a cautionary tale and a catalyst for just this kind of questioning. Roger Babson was an obdurate entrepreneur so giddy with the possibilities of the twentieth century that he saw even in gravity's downward tug an opportunity waiting to be leveraged. Born in 1875 in Gloucester, Massachusetts, Babson rejected the values of the town's "codfish aristocracy," instead holding fast to the good Protestant value of self-reliance. He turned his small investment newsletter into a trusted empire called Babson's Reports; founded Babson College near Wellesley, Massachusetts; and became both wealthy and famous when, it is said, he predicted the Great Crash of 1929 with his signature "Babson-chart."

Babson went barking into the post–World War II years with a beautiful idea: namely, that gravity, the most immutable law of the Enlightenment, was unacceptable and should be changed. "It seems as if there must be discovered some partial insulator of Gravity which could be used to save millions of lives and prevent accidents," he reasoned in his treatise "Gravity — Our Enemy Number One." His antipathy was steeped in grief. In the summer of 1893, Babson's three-year-old sister, Edith, had drowned in the Annisquam

River near their house. "Yes they say she was 'drowned,'" Babson wrote, but Edith was a fine swimmer. He blamed "Gravity which came up and seized her like a dragon and brought her to the bottom." In 1947 his teenage grandson had also drowned, after diving off a motorboat to rescue a friend. Less than eighteen months later, Babson formed the Gravity Research Foundation.

The foundation's underlying imperative was to learn all it could about gravity and defeat it. It rose to meet the gravity problem with seemingly unlimited funds and a fervor that suggests a near-total unawareness of its Sisyphean nature. Frozen-food magnate Clarence Birdseye, a Gravity Research Foundation trustee, suspected that a gravity insulator might be discovered by accident, through unrelated research. So 2,500 labs were contacted and asked to keep their eyes peeled. Three men were kept on permanent watch at the patent office. Thomas Edison once wondered aloud to Babson how it was that birds could fly — maybe there was something there. Thus a lavish collection of stuffed birds, from 5,000 different species, was amassed to be studied, just in case.

George Rideout — the foundation's president and Babson's longtime right-hand man — devised an annual essay contest with a $1,000 purse to inspire research on "the possibilities of discovering some partial insulator, reflector or absorber" of gravity. Babson and Rideout purchased a twenty-five-year run of *Time* magazine to use as an almanac, searching for correlations between international incidents and the phases of the moon. They wondered if gravity's pull on the body affected temperament and mailed an exploratory survey to subscribers of Babson's investment bulletin. And so straitlaced industrialists were asked to fill in their weight and to agree or disagree with such statements as "I love physical comfort," "I am an unimpressive talker," and "Ladies like me."

Gravity Aids for Weak Hearts, one of many pamphlets the foundation published, recommends lessening gravity's strain on the body by moving into a bungalow-style house or using a cane. *Gravity and Posture* states, "It behooves us therefore to give the body all possible aid in maintaining the proper gravity pull by wearing the right corset." Digging through what's left of the foundation's early affairs in the Babson College archives, I found an entire folder of these instructional guides, with such titles as *Gravity and the Weather* and *Gravity and Your Feet.* They linked gravity to the common cold, house fires, insomnia, poor crop conditions, tilted uteruses, the firing of General Douglas MacArthur, the shrinking of the elderly, tuberculosis, "worries," varicose veins, and hemorrhoids — which, one article asserts, "are merely varicose veins of the rectum."

"It was a time when amateurs could still hope and dream about being contributors," David Kaiser, a professor of physics and the history of science at MIT, told me. Kaiser has been fascinated with the Gravity Research Foundation ever since he wrote part of his doctoral thesis at Harvard about it. Babson's men were not experts, Kaiser explained, but their hearts were in the right place. "Babson thought, 'I'll just get all my buddies together and we'll fix it.'"

In some ways, undoing gravity seemed to be just another entrepreneurial project Babson felt compelled to get off the ground. (It was not, incidentally, his oddest venture: he also envisaged chocolate-covered fish to bail out Gloucester's economy; asbestos-lined pants pockets to prevent men from igniting their crotches when putting away their pipes; and, to right America's moral compass, the creation of a federal Department of Character Training, to be headed by the "Secretary of Character.") As a dedicated businessman, he couldn't allow gravity to go on wasting itself, pulling things in the same old direction for its own purposes and not ours. Electricity, light, magnetism — if all these things could be insulated and controlled, he argued, surely gravity could. It seemed only reasonable.

"World peace will come only as the Spirit of Jesus grows in the hearts of man and as the principles of birth control are taught to overcrowded nations and the latent power of gravity is used as freely as air, water and sunlight," Babson wrote in his autobiography, *Actions and Reactions*. He didn't know what caused gravity, and he didn't care. All he knew was that the damn stuff was everywhere, and only a sap would go on accepting its reprimand.

Inflation

"How did you get onto it?" Rainer Weiss, professor emeritus at MIT, wanted to know when I brought up the Gravity Research Foundation. "You got onto Mr. Chiao, so that's a bad sign," he joked. "And now you got onto this — which is an even worse sign!"

Weiss, a short, charismatic man of seventy-three, had returned from Louisiana late the previous night after visiting LIGO, the grand experiment he'd dreamed up thirty years earlier. "Observatory" is the key concept of the Laser Interferometer Gravitational-Wave Observatory. Ultimately, LIGO aims to diagram the universe by charting gravity waves emitted by moving bodies, just as we now see things by the light they emit. Mapping the "gravity sky" will open a new field of astronomy, Weiss said, one theoretically capable of assembling a picture of the big bang.

I'd hoped to find out more about LIGO in order to gauge just how farfetched Chiao's gravity-radio scheme actually is. Weiss resented my even comparing the two endeavors. He was a colleague of Chiao's at MIT for a time and still periodically writes Chiao to point out what he considers gravity radio's many unsolvable problems. "He says, 'Yeah, yeah. I'll get to it,'" Weiss said of Chiao. "I like him. He's a fine man. But he's not a doer. He likes to think about things."

On the phone, Weiss had urged me to stop writing about gravity waves altogether and cover "the real revolution" in physics — something called "inflation," which, he said, "is probably the most shocking discovery in my lifetime." Now that I'd flown across the country to meet Weiss, my first question was about Roger Babson. He squinted at me from behind his desk. "I'm worried about you," he said. Babson was a businessman, Weiss went on, who went loopy over gravity because his relative got killed in an airplane. I corrected him, relaying the fates of Babson's grandson and poor Edith. "And so gravity did the job on her too, eh?" Weiss said. He sounded sympathetic. Then his voice rose. "Well, there wouldn't have been any water there for her to swim around in if there was no gravity!"

I pointed to his office door in my defense. Weiss had taped up a flier announcing the Gravity Research Foundation's 2006 Awards for Essays on Gravitation, the fifty-seventh annual competition. The foundation has endured, its mission having matured from conquering gravity to understanding it. The essay contest, David Kaiser told me, is where graduate students now "look to see the coolest, hottest stuff." In his doctoral thesis, Kaiser argues that the contest helped return physicists to the study of gravity. Generations of graduate students had been told that there was no more work to do in the field after Einstein; one could merely plug numbers into his equations and futz around in the abstract. Rainer Weiss said that when he attended MIT in the fifties, general relativity was taught only in the math department. Gravity, in short, had lost its place in the physical world.

At that time, when a physics professor might earn only $5,000 a year, Babson's foundation was offering $1,000 for a brief theoretical essay on gravity. In 1953 a brilliant but down-on-his-luck postdoc named Bryce DeWitt submitted a paper and won. (DeWitt, with the backing of a southern financier Babson had introduced him to, went on to found an esteemed gravity-research institute at the University of North Carolina, Chapel Hill.) Two Princetonians took home the prize in 1954, and in 1957 it was a team from Cornell and Harvard. As the contest rules softened over time — it now solicits papers "on the subject of gravitation" — icons like Roger Penrose and Stephen Hawking won, too. In 1981 a young Berkeley professor named Raymond Chiao took second place for his essay "Gravitational Radiation Antenna." It was among Chiao's first staggering steps toward gravity radio — an idea, Chiao later told me, that initially leaped out at him from one of Bryce DeWitt's equations.

"Despite himself," Kaiser says, "Babson and his foundation — this band of misfits and amateurs — actually played a role in bringing great minds back to gravity. Fifty years ago no one was doing this work. Now it's what gets government funding. There are fancy conferences and big expensive equipment. It's what you make *NOVA* specials about."

The day before my meeting with Rainer Weiss, I tracked down a monument on the campus of Tufts University. I found the four-foot block of granite beside a chapel overlooking the Boston skyline.

THIS MONUMENT HAS BEEN
ERECTED BY THE
GRAVITY RESEARCH FOUNDATION
ROGER W. BABSON FOUNDER
IT IS TO REMIND STUDENTS OF
THE BLESSINGS FORTHCOMING
WHEN A SEMI-INSULATOR IS
DISCOVERED IN ORDER TO HARNESS
GRAVITY AS A FREE POWER
AND REDUCE AIRPLANE ACCIDENTS

Babson issued at least thirteen such monuments to various colleges, accompanying them in most cases with sizable gifts of stock. The endowments were to be held for a certain number of years and then dedicated to research in the fight against gravity. By the time the Tufts stock was freed up in 1989, it had appreciated to roughly half a million dollars. Absolved by the foundation's lawyers from its original antigravitational obligations, Tufts used this windfall to found its Institute of Cosmology, which is now a prestigious training ground for theoretical physicists. The institute's director, Alex Vilenkin, told me that new graduates are led to the monument and ordered to kneel so that an apple may be ceremoniously dropped on their heads.

I was beginning to see Babson's misdirected burst of energy as having loosed a ripple effect not unlike a gravity wave itself — nearly imperceptible but warping the fabric of legitimate physics ever so slightly wherever it reached. Still, I was unprepared for the epilogue to Vilenkin's story. "The funny thing is," he said, "we actually do work on antigravity."

He was referring to the theory of inflationary cosmology, the theoretical-physics revolution Weiss had urged me toward. The theory seeks to explain why the universe's expansion, once thought to be powered by momentum from the big bang, is, in fact, speeding up. Its answer is "dark energy," a power aggressively pushing everything in the cosmos away from everything else. It is, by definition, antigravity. Cosmologists now suspect that dark energy accounts for as much as 75 percent of the energy in existence; that is, our universe is mostly this thing we only just discovered. As Weiss put it, "It turns out there is no vacuum. The vacuum is full of stuff!"

Perhaps not since general relativity has a theory produced such maddeningly counterintuitive corollaries, calling into question fundamental presumptions of physics, even relativity itself. "A dark mystery is burbling up," Weiss told me, "and it says: we don't have a working knowledge of the universe."

Inflation happens to be David Kaiser's field. He was surprised but pleased that a scientist of Vilenkin's stature would refer to dark energy as antigravity. "Repulsive gravity," instead, has emerged as the term of art. Nothing about dark energy suggests it can be harnessed, generated, or, say, spread on the bottom of your shoe to facilitate expeditious slam-dunking. But physicists may worry that borrowing a shibboleth like antigravity would open them up to discussions with the wrong crowd.

I asked Kaiser how he'd respond to some earnest pseudoscientific hobbyist pointing to the relatively late discovery of dark energy to argue that the more fantastical antigravitational aspirations of laymen — of us all — should never be deemed impossible.

"I would grant that," he said. Though, he added, he wouldn't feel this same equanimity if he hadn't become so enamored of "our wacky friend Babson." When Kaiser does get mail like this, he chooses to be heartened by it. "Usually my reaction is, 'Isn't it great that people are interested in what we do all day?'" Besides, he told me, "crazy ideas are sometimes right."

Unwanted Degrees of Freedom

A low, charcoal-colored storm readied to break as I drove toward Yosemite to meet Raymond Chiao. In 2005 Chiao abandoned the flagship school of the University of California system — Berkeley — for its fledgling start-up, the University of California, Merced. The college was exactly three months old. It had eight hundred students. For the low-profile venture three hours' drive southeast of San Francisco, Chiao's appointment was "a major recruiting triumph." Merced's press release touted his research in gravitational radiation and listed his prize from the Gravity Research Foundation first among his many honors.

Chiao's laboratory is several miles from the unfinished campus, set back from Highway 99, past several cow yards, some sheep, and acres of blossoming almond orchards. The university had erected a lone building there on the decommissioned Castle Air Force Base. Nearby, midcentury dogfighters and transporters were set out to make an air museum. They crowded together on the flat, featureless land, a reliquary of ancient propeller planes in khaki and chrome.

Chiao had just finished lunch when I arrived. I asked after the gravity-radio prototype I'd read about — the paint cans, the wood. It was boxed up in a closet and forgotten, he said, fluttering his hands in front of him as though to exonerate himself. He had now arrived at a "much clearer and much simpler" concept. "I have reached a point of conclusion that is, I think, incontrovertible," he announced. "Unless I've made a mistake."

Gravity radio is essentially a transducer, Chiao explained. He would beam waves of electromagnetism, like those broadcast by a radio station, into two drops of helium, and they would bounce off the helium as gravity waves. The gravity waves would then be sent, however far and through whatever obstacles, to a second gravity radio, a receiver. There they would hit other drops of helium and be converted back to usable radio waves. Chiao's calculations led him to believe that nothing would be lost in these conversions. He could, in physics parlance, "freeze out" all the helium's "unwanted degrees of freedom." That is, if he cooled the helium, it would reflect one form of incoming energy by radiating it back as the other, and not in a disorganized array of other forms, such as sound or heat.

Chiao was trying to pick up exceedingly weak gravity waves through the disturbances they made in two exceedingly small drops of helium — as opposed to picking up very powerful waves, like those from exploding stars, through the disturbances they made in LIGO's 2.5-mile lasers. This, Chiao explained, looks impossible, given much of the field's thinking since Einstein. But to compensate for his antenna's infinitesimal size, Chiao was banking on triggering an elusive phenomenon of quantum mechanics. The helium, he believed, would act as a "superfluid": every atom would shudder under the gravity wave as one coherent object, amplifying the effect.[4]

"I don't care what people say," Chiao told me. "I really don't care. I think most people like Rai Weiss will say I'm a crackpot. But I know that I'm not, especially after writing this paper." Chiao had recently delivered a new paper on gravity radio to a colloquium in Snowbird, Utah, where he was awarded the prestigious Lamb Medal for unrelated work in optics. To celebrate, some former students had organized a series of laudatory panels called Chiaofest.

He was eager to walk me through the paper and printed out an even more recent version than the one he'd previously e-mailed. In this latest refinement, he'd linked his hypothesis to an ironclad constant in physics called Planck's mass. Although this maneuver did little to convince two other physicists I spoke to, it clearly put gravity radio on inviolable ground as far as Chiao was concerned. He underlined the paragraph in red, read it aloud slowly, then flipped the paper across his desk to me as if it were an unbeatable poker hand.

At Merced I'd expected to find a bitter man in exile. But Chiao seemed disarmingly content — relieved, even — to be there. He'd already put his start-up funds toward the sophisticated low-temperature lab that gravity radio would require, having ordered two European-built dilution refrigerators at a cost of half a million dollars. He was building from the ground up and would be ready to perform his experiment in five years. Maybe it wouldn't work. But the most important thing was to build his device and, through gravity radio, let the universe speak for itself. "In the end," Chiao said, "the truth will prevail. Especially in physics."

I'd heard, from more than one physicist, the real-life allegory of Joseph Weber, another accomplished Gravity Research Foundation honoree. In the late sixties, Weber announced he'd made the world's first successful detection of a gravity wave, using a stubby pair of aluminum bars in his University of Maryland lab. "It was a very unfortunate event for him," Rainer Weiss had explained, "because up to that time he'd been perfectly dispassionate about this." Physicists around the country, spurred to think seriously about gravity-wave experiments for the first time, eagerly built their own "Weber bars," trying to confirm his work and push it further. No one saw anything. But Weber clung to his story stubbornly for years. He traveled the country announcing his findings at conferences. Another physicist shadowed and heckled him to safeguard the integrity of the field. The two men had to be separated onstage at MIT after raising their fists. Demoralized but still adamant, Weber kept the bars running until he died in 2000.

When I asked Chiao how he knew he wasn't turning into another Joseph Weber, he characterized Weber's work as wishful thinking. "Allowing wishful thinking to dominate your assessment of reality is a sin," he said.

A Basement in Wellesley Hills

The Gravity Research Foundation is currently headquartered a mile from Babson College, in the suburb of Wellesley Hills. George Rideout Jr., the son of Babson's right-hand man, inherited the Gravity Research Foundation in 1988, six years before his father's death, and now administers its sole remaining activity, the essay contest, from a cluttered back room of his basement. Rideout, a subdued man with a long patrician chin, explained that the job is largely administrative. He photocopies contest submissions and mails them to an anonymous panel of judges every winter. Rideout minored in physics in college, he said, but cannot read the essays with any great understanding.

In preparation for our meeting, Rideout had set up a card table in the center of the basement headquarters, with a glass of water on a coaster resting in front of each of our seats. His setting also included a yellow legal pad, on top of which was written my name, followed by the date and a list of points he intended to cover. A battery of filing cabinets holding the complete archive of essay submissions lined one wall; nearby, hung among some crayon-drawn birthday cards, was a handwritten note from Abraham Lincoln to his surgeon general. Something Mr. Babson had collected, Rideout told me when I happened to notice it.

I had spent the previous day rifling through the Gravity Research Foundation's scant archives at Babson College, watching in those pages a man toil earnestly toward his own ham-fisted theory of everything — a man imagining a universe in which gravity was the greatest asset and the most pervasive menace. But I also detected in those papers the foundation's rising prestige as the years wore on. Then, following the minutes from Gravity Day 1958, the paper trail ran out.

Babson had sponsored Gravity Day every summer as part of a conference titled "Investments and Gravity" at the foundation's headquarters in New Hampshire. Attendees were largely from the business world. Thus a presentation on "Eliminating Weight" might dovetail with a talk like "Who Should Buy Mutual Funds?" Conventioneers were invited to see Isaac Newton's bed, view the stuffed-bird collection, or sit in special Japanese-made "Gravity Chairs," undulating wooden recliners that alleviated gravity's strain on the legs. "Remember," exclaimed the brochures, "that gravity is Enemy Number One for middle aged and older people."

Eventually some airline executives began attending, as did Igor Sikorsky, the inventor of the first successful helicopter. By 1958 the minutes depict a serious-minded crowd of 278 discussing general relativity and joking about various quack entrants in that year's essay contest. According to one attendee, many of the essayists had devised their arguments simply by reinventing gravity "from scratch, with a mind uncluttered by knowledge." None of the conventioneers pointed out that they were gathered there because ten years earlier, Roger Babson had done precisely that.

Babson, then an eighty-three-year-old man, was relegated to a brief paragraph at the end of the minutes. "Before the close of the session," the record concludes, "Mr. Babson reported on the question of the physical reality of the examples of levitation mentioned in the Bible." He'd polled Christian Scientists, Roman Catholics, and Protestant clergy and asked that the foundation convene every Easter to discuss the matter further.

George Rideout Jr. wasn't sure he had ever attended a Gravity Day. Much of what Rideout knew about Babson, he admitted, came from his father and the autobiography *Actions and Reactions*. But he had enjoyed romping around the New Hampshire compound as a child. He liked the Japanese Gravity Chairs; they were comfortable. "It's too bad," Rideout said from across the card table. "They sold that place, and I don't think they kept any of the chairs." He then pointed to what remained of the foundation's stuffed-bird collection. A long-necked specimen sat upright on the filing cabinets to my left, and another, squatter bird perched by the opposite wall. "It's some kind of duck up here," Rideout said, noting the webbed feet.

When I relayed what I'd learned at the Cosmology Institute at Tufts, Rideout reached behind him for a photo of the gravity monument. But his arm swept into his still full glass of water. The glass, having no choice, toppled over. Ice cubes skirted across the card table. There was suddenly a puddle. "Let me go get a towel," Rideout said.

A photocopy of "Black Holes Aren't Black," an essay by Stephen Hawking, had gotten the worst of the spill. I must not have noticed Rideout sliding it into position on the table as we spoke. He'd been bringing out various relevant artifacts as he steered our conversation down the talking points on his pad. Much of his presentation dwelled on the stature of the foundation today. I assumed he was saving the Hawking paper as a robust finale. It included hand-drawn diagrams, and I took its brusquely paradigm-shattering title as a sign of its significance in the history of physics. The Gravity Research Foundation awarded it third place in 1974.

Soon Rideout returned with a yellow hand towel, which he spread carefully across a dry section of the floor. He brought over the first few pages of the Hawking paper and arranged them on the towel. Then, with a kind of aloof but patient dignity, he knelt to pat them dry.

The Mystery Spot

After returning home from New England, my wife and I took a drive to the California redwood forest one Sunday afternoon. Our destination was the Mystery Spot, a small plot of land where "the laws of physics do not apply." "Within the Mystery Spot," a bright yellow brochure claims, "it appears as though every law of gravitation has gone haywire, turned topsy-turvy and just doesn't make sense." Grainy photos show a ball rolling uphill, someone leaning so far back he seems to hover, and an old man standing on a wall. According to the promotional materials, a number of theories have been worked up to try to explain this aberration: an excess of carbon dioxide or radiation, a "magma vortex," underground metal cones implanted by aliens.

A guide in a ranger-style uniform led a group of us up a hillside into a cabin he described as "the grandfather of all American funhouses." The wooden floor sloped in one direction and the walls skewed off in others. The guide announced that we were in an epicenter of mystery, where unknowable forces abound and the power of gravity frays. Then he began, slowly, to bend backward from the waist. He kept on bending — more, further, Matrix-like — until his hands touched down on the floor behind him. Emboldened, a boy there with his mother and older brother dove into a kind of handstand, corkscrewing his pubescent torso like a marlin snagged on an invisible line. It was stunning. I'd read about a psychology professor at UC Santa Cruz who brings his classes here to demonstrate how optical illusions operate. But I found myself ignoring any rational explanations of the Mystery Spot's mysteriousness, instead losing myself in a kind of simple glee. Even my wife, who had threatened to wait in the car, now seemed delighted by the strange push and mild nausea neither of us could deny feeling.

In 1940 a certain Mr. Prather first claimed to have discovered "puzzling variations in gravity" on this land and promptly opened the Mystery Spot as a kitschy "mind-boggling" amusement. It was an era of amusements and, not unconnectedly, of war. Science was moving in its one perfunctory direction: forward. Physics would build the bomb that gravity lowered on Japan with quiet and characteristic indifference, 32 feet per second, per second. Who then could fault Roger Babson for believing, with a faint and tender measure of desperation, that other trajectories were possible? "The harnessing of Gravity today is at the stage where the harnessing of electricity was when Ben Franklin flew his kite during a thunderstorm," Babson wrote in 1950. Free power from gravity "is the next thing on the scientists' agenda. It has been delayed by the Army's atomic craze to kill people; but it is coming," he insisted. "Be patient."

When I called Rainer Weiss a few weeks later to tell him about Chiao's move to Merced, he became suddenly optimistic. "I bet you something good will come of it," he said. Although he had no hope for gravity radio, he felt that a physicist of Chiao's caliber, given this new opportunity to do well-funded experiments, would end up making some important contribution to the field. Ideas are nice, Weiss explained, but they need to be checked and honed through actual experiments. In science, progress means industriously refining one's vision of the world until it reflects reality. This is to say that in science, progress means the exact opposite of what it meant for Roger Babson.

Babson assumed he could will the universe's most elemental mechanism into a more agreeable shape, that he could keep hammering away at it until it gave. In the closing chapter of *Actions and Reactions*, he wrote, "Perhaps the foremost lesson I have learned is that emotions rule the world, rather than statistics, information, or anything else." This may be his stoic confession, a recognition of the blinding force of his own impracticable idealism. He concedes that the longing for what should be possible does not easily give way in the implacable face of what actually is.

In the redwoods that afternoon, as gravity suddenly seemed to slacken its grip, I realized we'd been lured to the Mystery Spot by the promise of finding longing and reality finally, if only fleetingly, aligned. It felt like a celebration, a homecoming into our own imaginations. We were chattering, pointing, laughing. Suddenly a woman climbed onto a table, leaned startlingly far over the edge, and balanced there for a friend's camera. "This should be in a commercial," someone else yelled, overtaken by wonder.

I watched a little girl dressed entirely in pink, with pink sandals, pink fur trim on her coat, and pink-sequined fringe on her skirt, stretch out her arms and start to twirl on a curious axis. Then the tour guide cupped a hand to his mouth and hollered, "If anyone wants to walk up the walls, come with me!" And we followed him into the adjoining room, drifting uphill.

Notes

1. The theory of general relativity conceived of empty space as actually being a fabric of space and time, with gravity caused by the imprints objects make in it. Imagine the depression a basketball would make if placed in the center of a taut sheet; when a less massive object, like a marble, travels close to the ball, it will be derailed off its course and begin circling in the curved depression made by the ball. In this way, Earth orbits the sun, and we are forever foundering in the deep space-time trench around Earth. Einstein's gravity is not so much a force as a circumstance: the very material of the cosmos has crumpled steeply around you until, almost conspiratorially, all of your possible paths have been narrowed to one.

2. Gillette envisioned the universe as being composed of little masses that zipped around and collided. "In all the Cosmos there is naught but straight flying, bumping, caroming and again straight flying . . . A mass unit's career is but lumping, jumping, bumping, rejumping, rebumping, and finally unlumping . . . Gravitation," he concluded, "and backscrewing are synonymous."

3. Stray too far into the site's labyrinth of links, and you could end up, as I did, listening to an interview with "a man who saw somebody dear to him suddenly transform into a strange, reptilian creature." But with gravity most of us are ill equipped to distinguish between the hokum and the shreds of austere physics on which it is imaginatively based; "spooky action at a distance," for instance, is Einstein's phrase and describes a genuine conundrum of quantum physics.

4. For nearly a century, physics has lived with a kind of acute schizophrenia: general relativity and quantum mechanics, both useful predictors of the way things work in and of themselves, can't be made to fit together into a single "theory of everything." Quantum mechanics is used to understand phenomena on the atomic level and smaller, and relativity applies to those on much larger scales. The real significance of gravity radio, if it works, is that Chiao will have teased out an interface between the two.

IAN PARKER

Swingers

FROM *The New Yorker*

ON A SATURDAY EVENING a few months ago, a fundraiser was held in a downtown Manhattan yoga studio to benefit the bonobo, a species of African ape that is very similar to — but, some say, far nicer than — the chimpanzee. A flier for the event depicted a bonobo sitting in the crook of a tree, a superimposed guitar in its left hand, alongside the message SAVE THE HIPPIE CHIMPS! An audience of young, shoeless people sat cross-legged on a polished wooden floor, listening to Indian-accented music and eating snacks prepared by Bonobo's, a restaurant on Twenty-third Street that serves raw vegetarian food. According to the restaurant's takeout menu, "Wild bonobos are happy, pleasure-loving creatures whose lifestyle is dictated by instinct and Mother Nature."

The event was arranged by the Bonobo Conservation Initiative, an organization based in Washington, D.C., which works in the Democratic Republic of Congo to protect bonobo habitats and to combat illegal trading in bush meat. Sally Jewell Coxe, the group's founder and president, stood to make a short presentation. She showed slides of bonobos, including one captioned MAKE LOVE NOT WAR, and said that the apes, which she described as "bisexual," engaged in various kinds of sexual activity in order to defuse conflict and maintain a tranquil society. There was applause. "Bonobos are into peace and love and harmony," Coxe said, then joked, "They might even have been the first ape to discover marijuana." Images of bonobos were projected onto the wall behind her: they looked like chimpanzees but had longer hair, flatter faces, pinker lips, smaller ears, narrower bodies, and, one might say, more gravitas — a chimpanzee's arched brow looks goofy, but a bonobo's low, straight brow sets the face in what is easy to read as earnest contemplativeness.

I spoke to a tall man in his forties who went by the single name Wind and who had driven from his home in North Carolina to sing at the event. He was a musician and a former practitioner of "metaphysical counseling," which he also referred to as clairvoyance. He said that he had encountered bonobos a few years ago at Georgia State University, at the invitation of Sue Savage-Rumbaugh, a primatologist known for experiments that test the language-learning abilities of bonobos. (During one of Wind's several visits to GSU, Peter Gabriel, the British pop star, was also there; Gabriel played a keyboard, another keyboard was put in front of a bonobo, and Wind played flutes and a small drum.) Bonobos are remarkable, Wind told me, for being capable of "unconditional love." They were "tolerant, patient, forgiving, and supportive of one another." Chimps, by contrast, led brutish lives of "aggression, ego, and plotting." As for humans, they had some innate stock of bonobo temperament, but they too often behaved like chimps. (The chimp-bonobo division is strongly felt by devotees of the latter. Wind told me that he once wore a chimpanzee T-shirt to a bonobo event and "got shit for it.")

It was Wind's turn to perform. "Help Gaia and Gaia will help you," he chanted into a microphone, in a booming voice that made people jump. "Help bonobo and bonobo will help you."

In recent years, the bonobo has found a strange niche in the popular imagination, based largely on its reputation for peacefulness and promiscuity. The *Washington Post* recently described the species as copulating "incessantly"; the *Times* claimed that the bonobo "stands out from the chest-thumping masses as an example of amicability, sensitivity and, well, humaneness"; a PBS wildlife film began with the words "Where chimpanzees fight and murder, bonobos are peacemakers. And, unlike chimps, it's not the bonobo males but the females who have the power." The Kinsey Institute claims on its website that "every bonobo — female, male, infant, high or low status — seeks and responds to kisses." And in Los Angeles a sex adviser named Susan Block promotes what she calls "The Bonobo Way" on public-access television. (In brief: "Pleasure eases pain; good sex defuses tension; love lessens violence; you can't very well fight a war while you're having an orgasm.") In newspaper columns and on the Internet, bonobos are routinely described as creatures that shun violence and live in egalitarian or female-dominated communities; more rarely, they are said to avoid meat. These behaviors are thought to be somehow linked to their unquenchable sexual appetites, often expressed in the missionary position. And because the bonobo is the "closest relative" of humans, its comportment is said to instruct us in the fundamentals of human nature. To underscore the bonobo's status as a signpost species — a guide to human virtue, or at least modern dating — it is said to walk upright. (The *Encyclopaedia Britannica* depicts the species in a bipedal pose, like a chimpanzee in a sitcom.)

This pop image of the bonobo — equal parts dolphin, Dalai Lama, and Warren Beatty — has flourished largely in the absence of the animal itself, which was recognized as a species less than a century ago. Two hundred or so bonobos are kept in captivity around the world; but, despite being one of just four species of great ape, along with orangutans, gorillas, and chimpanzees, the wild bonobo has received comparatively little scientific scrutiny. It is one of the oddities of the bonobo world — and a source of frustration to some — that Frans de Waal, of Emory University, the high-profile Dutch primatologist and writer, who is the most frequently quoted authority on the species, has never seen a wild bonobo.

Attempts to study bonobos in their habitat began only in the 1970s, and those efforts have always been intermittent, because of geography and politics. Wild bonobos, which are endangered (estimates of their number range from 6,000 to 100,000), keep themselves out of view, in dense and inaccessible rain forests, and only in the Democratic Republic of Congo, where, in the past decade, more than 3 million people have died in civil and regional conflicts. For several years around the turn of the millennium, when fighting in Congo was at its most intense, field observation of bonobos came to a halt.

In recent years, however, some Congolese and overseas observers have returned to the forest and to the hot, damp work of sneaking up on reticent apes. The most prominent scientist among them is Gottfried Hohmann, a research associate at the Max Planck Institute for Evolutionary Anthropology, in Leipzig, Germany. He has been visiting Congo off and on since 1989. When I first called Hohmann, two years ago, he didn't immediately embrace the idea of taking a reporter on a field trip. But we continued to talk, and in the week after attending the bonobo fundraiser in New York I flew to meet Hohmann in Kinshasa, Congo's capital. A few days later, I was talking with him and two of his colleagues in the shade of an aircraft hangar in Kinshasa's airport for charter flights, waiting for a plane to fly us to the forest.

It was a hot morning. We sat on plastic garden chairs, looking out over a runway undisturbed by aircraft. The airport seemed half ruined. Families were living in one hangar, and laundry was hung to dry over makeshift shelters. A vender came by with local newspapers, which were filled with fears of renewed political violence. European embassies had been sending cautionary text messages to their resident nationals.

Hohmann is a lean, serious, blue-eyed man in his midfifties. He has a reputation for professional fortitude, but also for chilliness. One bonobo researcher told me that he was "very difficult to work with," and there were harsher judgments, too. He lives in Leipzig with Barbara Fruth, his wife and frequent scientific collaborator, and 'their three young children. Three or four times a year, he flies to Kinshasa, where he charters a light plane operated by an American-based missionary group. The plane takes him into the world's second-largest rain forest, in the Congo Basin, and puts him within hiking distance of a study site called Lui Kotal, where he has worked since 2002. When Hohmann first came to Congo — then Zaire — he operated from a site that could be reached only by sweating upriver for a week in a motorized canoe. "People think it's entertaining, but it's not," he told me, as we waited. "It's so slow. So hard." He added, "You always think there's going to be something round the next bend, but there never is." He is an orderly man who has learned how to withstand disorder, an impatient man who has reached some accommodation with endless delay.

Hohmann makes only short visits to Lui Kotal, but the camp is run in his absence by Congolese staff members on rotation from the nearest village and by foreign research students or volunteers. Two new camp recruits were joining Hohmann on this flight: Andrew Fowler, a tough-looking Londoner in his forties, was an experienced chimpanzee field worker with a Ph.D.; Ryan Matthews was a languid Canadian-American of thirty who had answered an online advertisement to be Lui Kotal's camp manager, for 300 euros a month. We had all met for the first time a few days earlier, in a café in the least lawless neighborhood of Kinshasa, where Hohmann had flatly noted that of all the overseas visitors he had invited to Lui Kotal over the years, only one had ever wanted to return. Fowler and Matthews were a bit wary of Hohmann, and so was I. We had exchanged small talk over a pink tablecloth, establishing, first, that the British say "bo-*noh*-bo"; Americans, "*bahn*-obo"; and Germans, something in between.

Fowler and Matthews had just taken their last shower before Christmas. They would be camping for at least nine months, detached from their previous lives except for access, once or twice a week, to brief e-mails. Fowler, emanating self-reliance, was impatient for the exile to come; he had brought little more than a penknife and a copy of *The Seven Pillars of Wisdom*. Matthews was carrying more. As we discovered over time, his equipment included a fur hat, a leather-bound photo album, an inflatable sofa, and goggles decorated with glitter. Matthews is a devotee of the annual Burning Man festival, in the Nevada desert, and this, apparently, had informed his African preparations.

Matthews would be keeping accounts and ordering supplies. Fowler's long-term plan was to find a postdoctoral research topic about bonobos, but his daily duty on this trip was to be a "habituator" — someone able to find the community of thirty or so bonobos known to live near the camp and to stay within sight of them as they moved from place to place, with the idea that future researchers might be able to observe them for more than a few seconds at a time. Fowler called it "chimp-bothering." (Watching bonobos, I understood, is not like ornithology; there's no pretense that you're not there.) It gave an insight into the pace of bonobo studies to realize that nearly five years after Hohmann first reached Lui Kotal, this process of habituation and identification — upon which serious research depends — remained unfinished.

"There's a satisfaction for a scientist to come home at night with his notebook filled," Hohmann said with a shrug. "The most happy people are always the ecologists. They go to the forest, and the trees are not running away." He and his colleagues were still "racing through the dark, trying to get IDs," and most of the interesting bonobo questions were still unanswered. Is male aggression kept in check by females? Why do females give birth only every five to seven years, despite frequent sexual activity? In the far distance, such lines of inquiry may converge at an understanding of bonobo evolution, Hohmann said, and, beyond, of the origins of human beings. "It's a long path, and because it's long, there are few people who do it. If it was quicker and easier? There are hundreds of people working with baboons and lemurs, so it's not so easy to find your niche. A student working with bonobos can close his eyes and pick a topic, and it *can't* be wrong."

We finally boarded a tiny plane. Our pilot was a middle-aged American with a straight back and a large mustache. As we took off, Matthews was speaking on a cell phone to his mother, in New Jersey — enjoying the final moments of reception before it was lost for the rest of the year. The Congo River was beneath us as we rose through patches of low clouds. Suddenly the plane seemed to fill with clouds, as if clouds were made of a dense white mist that could drift between airplane seats. The pilot turned to look — the fog seemed to be coming from the rear of the cabin — and then glanced at Hohmann, whose seat was alongside his. "Is that OK?" the pilot asked, in the most carefree tone imaginable. Hohmann said it was, explaining that liquid nitrogen, imported to freeze bonobo urine, must have been forced out of its canister by the change in air pressure. Meanwhile, Matthews told his mother, "The plane seems to be filling with smoke," at which point his phone dropped the call.

We flew inland, to the east. The Congo River looped away to the north. Bonobos live only south of the river. (Accordingly, they have been called "left-bank chimps.") The evolutionary tree looks like this: if the trunk is the common ape ancestor and the treetop is the present day, then the lowest — that is, the earliest — branch leads to the modern orangutan. That may have been about 16 million years ago. The next-highest branch, around 8 million years ago, leads to the gorilla; then, 6 million years ago, the human branch. The remaining branch divides once more, perhaps 2 million years ago. And this last split was presumably connected to a geographical separation: chimpanzees evolved north of the Congo River, bonobos to the south. Chimpanzees came to inhabit far-flung landscapes that had various tree densities; bonobos largely stayed in thick, gloomy forest. (Chimpanzees had to compete for resources with gorillas, but bonobos never saw another ape — one theory argues that this richer environment, by allowing bonobos to move and feed together as a leisurely group, led to the evolution of reduced rancor.) From the plane, we first looked down on a flat landscape of grassland dotted with patches of trees; this slowly became forest dotted with grassland patches; and then all we could see was a crush of trees barely making way for the occasional scribble of a Congo tributary.

After three hours, we landed at a dirt airstrip in a field of tall grass and taller termite mounds. There were no buildings in sight. We were just south of the equator, 500 miles from Kinshasa and 300 miles from the nearest road used by cars, in a part of the continent connected by waterways or by trails running through the forest from village to village, good for pedestrians and the occasional old bicycle. The plane left, and the airstrip's only infrastructure — a sunshade made of a sheet of blue plastic tied at each corner to a rough wooden post — was dismantled in seconds and taken away.

Joseph Etike, a quizzical-looking man in his thirties who is Hohmann's local manager, organized porters to carry our liquid nitrogen and our inflatable sofa. We first walked for an hour to Lompole, a village of thirty houses made of baked-earth bricks and thatched roofs, and stopped at Etike's home. "People were amazed when Gottfried first came to the village and asked about the bonobos," Etike recalled, standing beside his front door. (He spoke in French, his second language.) "They'd never heard of such a thing." His salary was reflected in his wardrobe: he was dressed in jeans and sneakers, while his neighbors wore flip-flops and battered shorts and Pokémon T-shirts. I asked Etike how local people had historically thought of the bonobo. "It depends on the family," he said. "In mine, there was a story that my great-great-grandfather became lost in the forest and was found by a bonobo, and it showed him the path. So my family never hunted them." But the tradition was somehow not fully impressed on Joseph as a boy, and when he was seventeen someone gave him bonobo meat, to his mother's regret. How did it taste? "Like antelope," he said. "No. Like elephant meat."

One afternoon in 1928, Harold Coolidge, a Harvard zoologist, was picking through a storage tray of ape bones in a museum near

Brussels. He examined a skull identified as belonging to a juvenile chimpanzee from the Belgian Congo and was surprised to see that the bones of the skull's dome were fused. In a young chimpanzee (and in a young human, too), these bones are not joined but can shift in relation to one another, like broken ice on a pond. He had to be holding an adult head, but it was not a chimpanzee's. Several similar skulls lay nearby.

Coolidge knew that this was an important discovery. But he was incautious; when the museum's director passed by, Coolidge mentioned the skull. The director, in turn, alerted Ernst Schwarz, a German anatomist who was already aware that there were differences between apes on either side of the Congo. And, as Coolidge later wrote, "In a flash Schwarz grabbed a pencil and paper," and published an article that named a new subspecies, *Pan satyrus paniscus,* or pygmy chimpanzee. This was the animal that eventually became known as the bonobo. (In fact, bonobos are barely smaller than chimpanzees, except for their heads, but Schwarz had seen only a head.) "I had been taxonomically scooped," Coolidge wrote. He had the lesser honor of elevating *Pan paniscus* to the status of full species, in 1933.

Live bonobos had already been seen outside Congo, but they, too, had been misidentified as chimps. At the turn of the century, the Antwerp zoo held at least one. Robert Yerkes, a founder of modern primatology, briefly owned a bonobo. In 1923 he bought two young apes and called one Chim and the other Panzee. In *Almost Human,* published two years later, he noted that they looked and behaved quite differently. Panzee was timid, dumb, and foul-tempered. "Her resentment and anger were readily aroused and she was quick to give them expression with hands and teeth," Yerkes wrote. Chim was a joy: equable and eager for new experiences. "Seldom daunted, he treated the mysteries of life as philosophically as any man." Moreover, he was a "genius." Yerkes's description, coupled with later study of Chim's remains, made it plain that he was *Pan paniscus:* bonobos had a good reputation even before they had a name. (Panzee was a chimpanzee; but, in defense of that species, her peevishness was probably connected to a tuberculosis infection.) Chim died in 1924, before his species was recognized.

For decades, "pygmy chimpanzee" remained the common term for these apes, even after "bonobo" was first proposed in a 1954 paper by Eduard Tratz, an Austrian zoologist, and Heinz Heck, the director of the Munich zoo. (They suggested, incorrectly, that "bonobo" was an indigenous word; they may have been led astray by Bolobo, a town on the south bank of the Congo River. In the area where Hohmann works, the species is called *edza*.) In the thirties, that zoo had three members of *Pan paniscus,* and Heck and Tratz had studied them. By the time their paper, the first based on detailed observations of bonobo behavior, was published, the specimens were dead, allegedly killed by stress during Allied air raids. (The deaths have been cited as evidence of a bonobo's innate sensitivity; the zoo's brute chimpanzees survived.) As Frans de Waal has noted, Heck and Tratz's pioneering insights — they wrote that bonobos were less violent than chimps, for example — did not become general scientific knowledge and had to be rediscovered.

Twenty years passed before anyone attempted to study bonobos in the wild. In 1972 Arthur Horn, a doctoral candidate in physical anthropology at Yale, was encouraged by his department to travel alone to Zaire; on the shore of Lake Tumba, 300 miles northwest of Kinshasa, he embarked on the first bonobo field study. "The idea was to gather all the information about how bonobos lived, what they did — something like Jane Goodall," Horn told me. Goodall was already famous for her long-term study of chimpanzees in Gombe, Tanzania, and for her poise in the films made about her by the National Geographic Society and others. Thanks in part to her work, the chimpanzee had taken on the role of model species for humans — the instructive nearest neighbor, the best living hint of our past and our potential. (That role had previously been held, at different times, by the gorilla and the savanna baboon.) At this time, Goodall had confidence that chimpanzees were "by and large, rather 'nicer' than us."

Horn's attempt to follow Goodall's model was thwarted. He spent two years in Africa, during which time he observed bonobos for a total of about six hours. "And, when I did see them, as soon as they saw me they were gone," he told me.

In 1974, not long after Horn left Africa, Goodall witnessed the start of what she came to call the Four-Year War in Gombe. A chimpanzee population split into two, and over time one group wiped out the other, in gory episodes of territorial attack and cannibalism. Chimp aggression was already recognized by science, but chimp warfare was not. "I struggled to come to terms with this new knowledge," Goodall later wrote. She would wake in the night, haunted by the memory of witnessing a female chimpanzee gorging on the flesh of an infant, "her mouth smeared with blood like some grotesque vampire from the legends of childhood."

Reports of this behavior found a place in a long-running debate about the fundamentals of human nature — a debate, in short, about whether people were nasty or nice. Were humans savage but for the constructs of civil society (Thomas Hobbes)? Or were they civil but for the corruptions of society (Jean-Jacques Rousseau)? It had not taken warring chimps to suggest some element of biological inheritance in human behavior, including aggression: the case had been made, in its most popular recent form, by Desmond Morris, in *The Naked Ape,* his 1967 bestseller. But if chimpanzees had once pointed the way toward a tetchy but less than menacing common ancestor, they could no longer do so: Goodall had documented bloodlust in our closest relative. According to Richard Wrangham, a primatologist at Harvard and the author, with Dale Peterson, of *Demonic Males* (1996), the Gombe killings "made credible the idea that our warring tendencies go back into our prehuman past. They made us a little less special."

Meanwhile, bonobo studies began to gain momentum. Other scientists followed Horn into the Congo Basin, and they set up two primary field sites. One, at Lomako, 300 miles northeast of Lake Tumba, came to be used by Randall Susman, of Stony Brook University, and his students. Farther to the east, Takayoshi Kano, of Kyoto University, in Japan, made a survey of bonobo habitats on foot and on bicycle, and in 1974 he set up a site at the edge of a village called Wamba. Early data from Wamba became better known than Lomako's: the Japanese spent more time at their site and saw more bonobos. Susman, however, can take credit for the first bonobo book: he edited a collection of papers given at the first bonobo symposium, in Atlanta, in 1982.

In the winter of 1983–84, in an exploration that was less grueling but as influential as any field research, Frans de Waal turned his attention from chimps to bonobos and spent several months observing and videotaping ten bonobos in the San Diego Zoo. He had recently published *Chimpanzee Politics: Power and Sex Among Apes* (1982), to great acclaim, and, as de Waal recently recalled, "Most

people I talked to at the time would say, 'Why would you do bonobos if you can do real, big chimpanzees?'" Among the papers that
drew on his studies in San Diego, one was particularly noticed in
the academy. In "Tension Regulation and Nonreproductive Functions of Sex in Captive Bonobos," de Waal reported that these apes
seemed to be having more sex, and more kinds of sex, than was really necessary. He recorded 17 brief episodes of oral sex and 420
equally brief episodes of face-to-face mounting. He also saw 43 instances of kissing, some involving "extensive tongue-tongue contact."

In the late 1980s, Gottfried Hohmann was an ambitious scientist in
his thirties; he spent nearly three years in southern India research-
ing vocal communication in macaques and langurs. "But it was difficult then to get funding for India," Hohmann told me. "And the
bonobo thing was just heating up. Frans's paper really affected
everyone" in the scientific community. "Tongue-kissing apes? You
can't come up with a better story. Then people said to me, 'We
want you to go in the field.'" Hohmann ran his hand back and
forth over his head. "So," he said.

   We were sitting on a wooden bench at the edge of a forest clearing barely larger than a basketball court, talking against a constant
screech — an insect tinnitus that the ear never quite processed
into silence. Trees rose a hundred feet all around, giving the impression that we had fallen to the bottom of a well. Two days after
our plane touched down, we had reached Lui Kotal. In the inter-
vening hours, which were inarguably more challenging for the
three newcomers than for Hohmann, we had first camped in a violent rainstorm, then followed some unflagging porters on a trail
that led through the hot, soupy air of the forest and along waist-
high streams that flowed over mud. We had then camped again before crossing a fast-flowing river in an unsteady canoe.

   Now, at five in the afternoon, the light at Lui Kotal was beginning to fade. People who work there make do with little sun — and
with a horizon that is directly overhead. Around us, the wall of vegetation was solid except where broken by paths: one led back to the
village; another led into that part of the forest where Hohmann
and his team have permission to roam — an area, six miles by five,
whose boundaries are streams and rivers. In the clearing stood a

dozen structures with thatched roofs and no walls. Some of these sheltered tents; a larger one was a kitchen, where an open fire was burning; and another was built over a long wooden table, beside which hung a 2006 Audubon Society calendar that had been neatly converted — with glue, paper, and an extravagant superfluity of time — into a 2007 calendar. At the table sat two young American volunteers who were not many weeks away from seeing the calendar's images repeat. Pale, skinny men in their twenties, wearing wild beards, they looked as if they needed rescuing from kidnappers. Three others were less feral and had been in the camp for a shorter time: a young British woman volunteer, an Austrian woman who had recently graduated from the University of Vienna, and a Swiss Ph.D. student attached to the Max Planck Institute.

Hohmann, shirtless, was in an easy mood, knowing that much of the logistical and political business of the trip was now done. Before leaving the village, I'd seen something of a bonobo researcher's extended duties. The men of Lompole had convened around him, their arms crossed and hands tucked into their armpits. Hohmann remained seated and silent as an angry debate began — as Hohmann described it, between villagers who were unhappy about the original deal that compensated the village for having to stop hunting around Lui Kotal (this had involved a bulk gift of corrugated iron, to be used for roofs) and those who worked directly for the project and saw the greater advantage in stability and employment. Hohmann had finally gotten up and delivered a forceful speech in Lingala, Congo's national language. He finished with a moment of theater: he loomed over his main antagonist, wagging his finger. "It's good to remind him now and then how short he is," Hohmann later said, smiling.

By 1989, Hohmann told me, he had read enough bonobo literature to be tempted to visit Zaire. Even if one left aside French kissing, he said, "the bonobo allured me. I thought, *This* is a species." By then, thanks to field and captive studies, a picture of bonobo society had begun to emerge, and some peculiar chimpanzee-bonobo dichotomies had been described. Besides looking and sounding different from chimpanzees (bonobos let out high whoops that can seem restrained alongside chimpanzee yelling), bonobos seemed to order their lives without the hierarchical fury and violence of chimpanzees. ("With bonobos, everything is peaceful," Takeshi Furuichi, a Japanese researcher who worked with Takayoshi Kano at Wamba, told me. "When I see bonobos, they seem to be enjoying their lives. When I see chimpanzees, I am very, very sorry for them, especially for the high-ranking males. They really have to pay attention.") In captivity, at least, male bonobos never ganged up on females, although the reverse sometimes occurred. The bonds among females seemed to be stronger than among male chimpanzees, and this was perhaps reinforced by sexual activity, by momentary episodes of frottage that bonobo experts refer to as "genito-genital rubbing," or "g-g rubbing." And, unusually, the females were said to be sexually receptive to males even at times when there was no chance of conception.

"We said, 'We have to answer: Why is it like this?'" Hohmann said. "The males, the physically superior animals, do not dominate the females, the inferior animals? The males, the genetically closely related part of any bonobo group, do not cooperate, but the females, who are not related, do cooperate? It is not only different from chimpanzees but it violates the rules of social ecology."

Hohmann flew to Zaire and eventually set up a small camp in Lomako forest, a few miles from the original Stony Brook site. His memory is that Susman's camp had been unused for years, but Susman told me that it was still active and that Hohmann was graceless in the way that he took over the forest. And although Hohmann said that he worked with a new community of bonobos, Susman said that Hohmann inherited bonobos that were already habituated and failed to acknowledge this research advantage. Whatever the truth, the distrust seems typical of the field. The challenges of bonobo research call for chimpanzee vigor, and this leads to animosities. Susman told me that Hohmann was the kind of man who, "if he was sitting by the side of the road and needed a filter for his Land Rover, people would drive right by. Even if they had five extra filters in the trunk."

When a researcher has access to a species about which little is known, and whose every gesture seems to echo a human gesture, and whose eyes meet a human gaze, there is a temptation simply to stare until you have seen enough to tell a story. That is how Hohmann judged the work of Dian Fossey, who made long-term observations of gorillas in Rwanda, and the work of Jane Goodall, at least at the start of her career. "They lived with the apes and for the apes," he said. "It was 'Let's see what I'm going to get. I enjoy it anyway, so whatever I get is fine.'" And this is how Hohmann regarded the Japanese researchers, for all their perseverance. The Wamba site produced a lot of data on social and sexual relations, and Kano published a book about bonobos, which concluded with the suggestion that bonobos illuminated the evolution of human love. But "what the Japanese produced was not really satisfying," Hohmann said. "It was narrative and descriptive. They are not setting out with a question. They want to *understand* bonobos." Moreover, the Japanese initially lured bonobos with food, as Goodall had lured chimpanzees. This was more than habituation. At Wamba, bonobos ate sugar cane at a field planted for them. The primatological term is "provisioning"; Hohmann calls it opening a restaurant. (As an example of the possibly distorting impact of provisioning, Hohmann noted that the Wamba females had far shorter intervals between births than those at Lomako.)

Hohmann's first stay at Lomako lasted thirteen months. Halfway through, Barbara Fruth, a German Ph.D. student, flew to join him; they eventually married. (Up until then, "I was not thinking of having a family," Hohmann said. "I was just doing what I did. I said, 'I don't have the time, and who's crazy enough to join me?'") Hohmann and Fruth flew back and forth between Germany and Lomako, and the bonobos eventually became so habituated that they would sometimes fall asleep in front of their observers. The Max Planck Institute is not a university; it supports an academic life that many professors elsewhere would find enviable, one of long-term funding and no undergraduates. Hohmann was able to publish slowly. Though not immune to the charms of ape-watching, he was at pains to set himself precise research goals. How did bonobos build nests? How did they share food? As one of his colleagues described it, Hohmann wanted to avoid being dirtied by the stain of primatology — a discipline regarded by some in biology as being afflicted by personality cults and overextrapolation. The big bonobo picture might one day emerge, but it would happen only after the rigorous testing of hypotheses in the forest. When a publisher asked Hohmann for a bonobo book, he responded that it was too soon. "Gottfried's one of those people who don't want to risk being criticized, so they make absolutely certain that they've completely nailed everything down before they publish," Richard Wrangham told me, with a mixture of respect and impatience.

In 1997, not long after the birth of their first child, Hohmann and Fruth decided to live in Congo full time. They leased a house in Basankusu, the nearest town to Lomako with an airstrip. Hohmann had already picked up the keys when civil war intervened. The troops of Laurent Kabila, the rebel leader and future president, were at that time making a long traverse from west to east — they eventually reached Kinshasa, and President Mobutu Sese Seko fled. One day, when Hohmann was at Lomako without his wife, soldiers from the government side turned up and gave him a day to leave. "They wanted to get everyone out of the area who might help the rebels," Hohmann said. (Around the same time, the Japanese researchers abandoned Wamba.) Hohmann took only what he could carry. On his way back to Kinshasa, he was interrogated as a suspected spy.

The bonobo fell out of view of scientists at the very moment that the public discovered an interest. In 1991 *National Geographic* sent Frans Lanting, a Dutch photographer, to photograph bonobos at Wamba. "At the time, there were no pictures of bonobos in the wild," Lanting recently told me. "Or, at least, no professional documentation." On his assignment, Lanting contracted cerebral malaria. But he was stirred by his encounter with the bonobos. "I became sure that the boundaries between apes and humans were very fluid," he said. "You can't call them animals. I prefer 'creatures.' It was haunting, the way they knew as much about you as you knew about them." It became his task, he later told Frans de Waal, "to show how close we are to bonobos, and they to us."

Many of his photographs were sexually explicit. "*National Geographic* found the pictures of sexuality hard to bear," Lanting said. "That was a place the magazine was not ready to go." The magazine printed only tame images. Not long after, Lanting contacted de Waal, who had recently taken up a post at Emory as a professor of primate behavior and a researcher at the Yerkes National Primate Research Center. Agreeing to collaborate, they approached the German magazine *Geo*. As de Waal recently told me, laughing, "Naturally, *Geo* put two copulating bonobos on the cover." Not long afterward, *Scientific American* printed an illustrated article. In 1997 the Dutchmen brought out a handsome illustrated book, *Bonobo: The Forgotten Ape*.

By this time, the experiments of Sue Savage-Rumbaugh had

drawn the public's attention to Kanzi, a bonobo said to be unusually skilled at communicating with humans. (Savage-Rumbaugh's claims for Kanzi have been a source of controversy among linguists.) But de Waal's book established the reputation of the species in the mass media. Lanting's photographs, which have been widely republished, showed bonobos lounging at Wamba's sugar cane field, trying yoga stretches, and engaging in various kinds of sexual contact. A few pictures showed bonobos up on two feet. (As a caption noted, these upright bonobos were handling something edible and out of the ordinary — cut sugar cane, for example — suggesting a pose dictated by avidity, like a man bent over a table in a pie-eating contest.) In his text, de Waal interviewed field researchers, including Hohmann, and was fastidious at the level of historical and scientific detail. But his rhetoric was richly flavored and emphasized a sharp contrast between bonobos and chimpanzees. "The chimpanzee resolves sexual issues with power," he wrote. "The bonobo resolves power issues with sex." ("If chimpanzees are from Mars, bonobos must be from Venus," de Waal wrote on a later occasion.) Bonobos were more "elegant" than chimpanzees, he said, and their backs appeared to straighten "better" than those of chimpanzees: "Even chimpanzees would have to admit that bonobos have more style."

In a recent conversation, de Waal told me, "The bonobo is female-dominated, doesn't have warfare, doesn't have hunting. And it has all this sex going on, which is problematic to talk about — it's almost as if people wanted to shove the bonobo under the table." *The Forgotten Ape* presented itself as a European tonic to American prudishness and the vested interests of chimpanzee scientists. The bonobo was gentle, horny, and — de Waal did not quite say it — Dutch. Bonobos, he argued, had been neglected by science because they inspired embarrassment. They were "sexy," de Waal wrote (he often uses that word where others might say "sexual"), and they challenged established, bloody accounts of human origins. The bonobo was no less a relative of humans than the chimpanzee, de Waal noted, and its behavior was bound to overthrow "established notions about where we came from and what our behavioral potential is."

Though de Waal stopped short of placing bonobos in a state of blissful serenity (he acknowledged a degree of bonobo aggression), he certainly left a reader thinking that these animals knew how to live. He wrote, "Who could have imagined a close relative of ours in which female alliances intimidate males, sexual behavior is as rich as ours, different groups do not fight but mingle, mothers take on a central role, and the greatest intellectual achievement is not tool use but sensitivity to others?"

The appeal of de Waal's vision is obvious. Where, at the end of the twentieth century, could an optimist turn for reassurance about the foundations of human nature? The sixties were over. Goodall's chimpanzees had gone to war. Scholars such as Lawrence Keeley, the author of *War Before Civilization* (1996), were excavating the role of warfare in our prehistoric past. And as Wrangham and Peterson noted in *Demonic Males*, various nonindustrialized societies that were once seen as intrinsically peaceful had come to disappoint. Margaret Mead's 1928 account of a South Pacific idyll, *Coming of Age in Samoa,* had been largely debunked by Derek Freeman in 1983. The people identified as "the Gentle Tasaday" — the Philippine forest-dwellers made famous in part by Charles Lindbergh — had been redrawn as a small, odd community rather than as an isolated ancient tribe whose mores were illustrative. *The Harmless People,* as Elizabeth Marshall Thomas referred to the hunter-gatherers she studied in southern Africa, had turned out to have a murder rate higher than that in any American city. Although the picture was by no means accepted universally, it had become possible to see a clear line of thuggery from ape ancestry to human prehistory and on to Srebrenica. But if de Waal's findings were true, there was at least a hint of respite from the idea of ineluctable human aggression. If chimpanzees are from Hobbes, bonobos must be from Rousseau.

De Waal, who was described by *Time* in 2007 as one of the hundred influential people who "shape our world," effectively became the champion — soft-spoken, baggy-eyed, and mustachioed — of what he called the "hippies of the primate world," in lectures and interviews, and in subsequent books. In *Our Inner Ape: A Leading Primatologist Explains Why We Are Who We Are* (2005), he wrote that bonobos and chimpanzees were "as different as night and day." There had been, perhaps, a vacancy for a Jane Goodall figure to represent the bonobo in the broader culture, but neither Hohmann nor Kano had occupied it; Hohmann was too dour, and

Kano was not fluent in English. Besides, the bonobo was beyond the reach of all but the most determined and best-financed television crew. After 1997 that Goodall role — at least in a reduced form — fell to de Waal, though his research was limited to bonobos in captivity. At the time of the book's publication, de Waal told me, he could sense that not everyone in the world of bonobo research was thrilled for him, "even though I think I did a lot of good for their work. I respect the field workers for what they do, but they're not the best communicators." He laughed. "Someone had to do it. I have cordial relationships with almost all of them, but there were some hard feelings. It was 'Why is he doing this and why am I *not* doing this?'"

De Waal went on, "People have taken off with the word 'bonobo,' and that's fine with me" — although he acknowledged that the identification has sometimes been excessive. "Those who learn about bonobos fall too much in love, like in the gay or feminist community. All of a sudden, here we have a politically correct primate, at which point I have to get into the opposite role and calm them down: bonobos are not *always* nice to each other."

At the Lui Kotal camp, which Hohmann started five years after being expelled from Lomako, the people who were not tracking apes spent the morning under the Audubon calendar as the temperature and the humidity rose. Ryan Matthews put out solar panels to charge a car battery powering a laptop that dispatched e-mail through an uncertain satellite connection. Or, in a storage hut, he arranged precious cans of sardines into a supermarket pyramid. We sometimes heard the sneezelike call of a black mangabey monkey. For lunch we ate cassava in its local form, a long, cold, gray tube of boiled dough — a single *gnocco* grown to the size of a dachshund. A radio brought news of gunfire and rocket attacks in Kinshasa: Jean-Pierre Bemba, the defeated opposition candidate in the previous year's presidential elections, had ignored a deadline to disarm his militia, and hundreds had been killed in street fighting. The airport that we had used had been attacked. The Congolese camp members — including, at any time, two bonobo field workers, a cook, an assistant cook, and a fisherman, working on commission — were largely pro-Bemba or at least antigovernment, a view expressed at times as nostalgia for the rule of Mobutu Sese

Seko. Once they sang a celebratory Mobutu song that they had
learned as schoolchildren.

"It was so easy for Frans to charm everyone," Hohmann said of
de Waal one afternoon. "He had the big stories. We don't have
the big stories. Often we have to say, 'No, bonobos can be terribly
boring. Watch a bonobo and there are days when you don't see
anything — just sleeping and eating and defecating. There's no
sex, there's no food sharing.'" During our first days in camp, the
bonobos had been elusive. "Right now, bonobos are not vocaliz-
ing," Hohmann said. "They're just there. And if you go to a zoo, if
you give them some food, there's a frenzy. It's so different."

Captivity can have a striking impact on animal behavior. As Craig
Stanford, a primatologist at the University of Southern California,
recently put it, "Stuck together, bored out of their minds — what is
there to do except eat and have sex?" De Waal has argued that even
if captive bonobo behavior is somewhat skewed, it can still be use-
fully contrasted with the behavior of captive chimpanzees; he has
even written that "only captive studies control for environmental
conditions and thereby provide conclusive data on interspecific
differences." Stanford's reply is that "different animals respond
very differently to captivity."

In the wild, bonobos live in communities of a few dozen. They
move around in smaller groups during the day, in the pattern of a
bus-tour group let loose at a tourist attraction, then gather together each night to build new treetop nests of bent and half-
broken branches. But they stay in the same neighborhood for a lifetime. When Hohmann found bonobos on his first visit to Lui Kotal,
he could be confident that he would find the same animals in subsequent years. On this trip the bonobos had been seen, but they
were keeping to the very farthest end of Hohmann's 20,000-acre
slice of forest: a two-hour walk away. ("They are just so beautiful,"
Andrew Fowler, the British habituator, said after seeing them for
the first time. "I can't put it any other way.") There was talk of setting up a satellite camp at that end — a couple of tents in a small
clearing — but weighing against the plan was the apparently serious risk of attack by elephants. (Forest elephants headed an impressive lineup of local terrors, above leopards, falling trees, driver
ants, and the green mambas that were sometimes seen on forest
paths.) So the existing arrangement continued: two or three peo-

ple would go into the forest and hope to follow bonobos to their nest site at night; the following day, two or three others would reach that same point before dawn.

When I went out one morning with Hohmann and Martin Surbeck, the Swiss Ph.D. student, the hike began at quarter to four, and there were stars in the sky. We walked on a springy path — layers of decaying leaves on sand. I wore a head torch that lit up thick, atticlike dust and, at one moment, a bat that flew into my face. We stepped over fallen tree trunks in various states of decay, which sprouted different kinds of fungi; after an hour or so, we reached one on which local poachers had carved a graffiti message. Poachers, whose smoked bonobo carcasses can fetch five dollars each in Kinshasa's markets, have often been seen in the forest, and their gunfire often heard. Their livelihood was disrupted last year when Jonas Eriksson, a Swedish researcher on a visit to Lui Kotal, burned down their forest encampment. I was later given a translation of the graffiti: JONAS: VAGINA OF YOUR MOTHER.

Hohmann stopped walking at half past five, at a point he knew to be within a few hundred feet of where the bonobos had nested. Bonobos sleep on their backs — "maybe holding to a branch with just one foot, and the rest of the body looking very relaxed," Hohmann had said, adding that "nest building is the only thing that sets great apes aside from all other primates." (He speculates that the REM-rich sleep that nests allow may have contributed to the evolution of big brains.) We would hear the bonobos when they woke. When we turned off our flashlights, there was a hint of light in the sky, enough to illuminate Surbeck using garden clippers to cut a branch from a tree and snip it into a Y shape about four feet long; he tied a black plastic bag across the forked end, to create a tool that hinted at a lacrosse stick but was designed to catch bonobo urine as it dripped from treetops. Surbeck's dissertation was on male behavior: he would measure testosterone levels in the urine of various bonobos in the hope that power structures not easily detected by observation would reveal themselves. (If an evidently high-ranking male had relatively low testosterone, for example, that might say something about the power he was drawing from his mother. A male bonobo typically has a lifelong alliance with his mother.)

There was a rustle of leaves in the high branches, like a downdraft of wind. To walk toward the sound, we had to leave the trail, and Surbeck cut a path through the undergrowth, again using clippers, which allowed for progress that was quieter, if less cinematic, than a swinging machete. We stopped after a few minutes. I looked upward through binoculars and, not long afterward, removed the lens caps. The half-light reduced the forest to blacks and dark greens, but a hundred feet up I could see a bonobo sitting silently in the fork of a branch. Its black fur had an acrylic sheen. It was eating the tree's small, hard fruit; as it chewed, it let the casing of each fruit fall from the corner of its mouth. The debris from this and other bonobos dropped onto dead leaves on the forest floor, making the sound of a rain shower just getting under way.

In the same tree, a skinny bonobo infant walked a few feet from its mother, then returned and clambered, wriggling, into the mother's arms — and then did the same thing again. And there were glimpses, through branches, of other unhurried bonobos as they scratched a knee or glanced down at us, unimpressed, or stretched themselves out like artists' models. Hohmann had plucked a large, rattling leaf from a forest-floor shrub that forms a key part of the bonobo diet, and he began to shred it slowly, as if eating it: bonobo researchers aim to present themselves as animals nonchalantly feeding rather than creepily stalking. He and Surbeck made solemn, urgent notes in their waterproof notebooks and whispered to one another. They were by now aware of some twenty bonobos above us and could identify many by name (Olga, Paulo, Camillo). A fact not emphasized in wildlife films is that ape identification is frequently done by zoomed-in inspection of genitals. A lot of the conversation at Lui Kotal's dinner table dealt with scrotal shading or the shape of a female bonobo's pink sexual swelling. ("This one is like chewing gum spit out," Caroline Deimel, the Austrian, once said of a female.)

At about six-thirty, the bonobos started moving down the trees — not with monkey abandon but branch by branch, with a final thud as they dropped onto the forest floor. Then they walked away on all fours, looking far tougher — and more lean and muscular — than any zoo bonobo. An infant lay spread-eagled on the back of its mother in a posture that the scientific literature sweetly describes as "jockey style." (A bonobo's arms are shorter than a chimpanzee's, and its back is horizontal when it walks. A chimpanzee slopes to the rear.) As the last of the bonobos strolled off, we lost sight of them: the undergrowth stopped our view at a few feet. We walked in the direction they seemed to have gone and hoped to hear a call or the sound of moving branches. Hohmann told me that bonobos sometimes gave away their position by flatulence. The forest was by now hot and looked like a display captioned SNAKES in a natural history museum: plants pulled at our clothes, trees crumbled to dust, and the ground gave way to mud.

We heard a sudden high screech ahead — *whah, whah!* — and then saw, coming back in our direction, a reddish blur immediately followed by black. We heard the gallop of hands and feet on the ground, and a squeal. Hohmann told me in a whisper that we had seen a rare thing — a bonobo in pursuit of a duiker, a tiny antelope. "We were very close to seeing hunting," he said. "Very close." The bonobo had lost the race, Hohmann said, but if it had laid a hand on the duiker in its first lunge the results would have been bloody. Hohmann has witnessed a number of kills, and the dismembering, nearly always by females, that follows. Bonobos start with the abdomen; they eat the intestines first, in a process that can leave a duiker alive for a long while after it has been captured.

For a purportedly peaceful animal, a bonobo can be surprisingly intemperate. Jeroen Stevens is a young Belgian biologist who has spent thousands of hours studying captive bonobos in European zoos. I met him last year at the Planckendael Zoo, near Antwerp. "I once saw five female bonobos attack a male in Apenheul, in Holland," he said. "They were gnawing on his toes. I'd already seen bonobos with digits missing, but I'd thought they would have been bitten off like a dog would bite. But they really *chew*. There was flesh between their teeth. Now that's something to counter the idea of" — Stevens used a high, mocking voice — "'Oh, I'm a bonobo, and I love *everyone*.'"

Stevens went on to recall a bonobo in the Stuttgart Zoo whose penis had been bitten off by a female. (He might also have mentioned keepers at the Columbus and San Diego zoos who both lost bits of fingers. In the latter instance, the local paper's generous headline was APE RETURNS FINGERTIP TO KEEPER.) "Zoos don't know what to do," Stevens said. "They, too, believe that bonobos are less aggressive than chimps, which is why zoos want to *have* them. But as soon as you have a group of bonobos, after a while you have this really violent aggression. I think if zoos had bonobos in big enough groups" — more like wild bonobos — "you would even see them killing." In Stevens's opinion, bonobos are "very tense. People usually say they're relaxed. I find the opposite. Chimps are more laid-back. But if I say I like chimps more than I like bonobos, my colleagues think I'm crazy."

At Lui Kotal, not long after we had followed the bonobos for half a day and seen a duiker run for its life, Hohmann recalled what he described as a "murder story." A few years ago, he said, he was watching a young female bonobo sitting on a branch with its baby. A male, perhaps the father of the baby, jumped onto the branch in apparent provocation. The female lunged at the male, which fell to the ground. Other females jumped down onto the male, in a scene of frenzied violence. "It went on for thirty minutes," Hohmann said. "It was terribly scary. We didn't know what was going to happen. Shrieking all the time. Just bonobos on the ground. After thirty minutes, they all went back up into the tree. It was hard to recognize them, their hair all on end and their faces changed. They were really different." Hohmann said that he had looked closely at the scene of the attack, where the vegetation had been torn and flattened. "We saw fur, but no skin, and no blood. And he was gone." During the following year, Hohmann and his colleagues tried to find the male, but he was not seen again. Although Hohmann has never published an account of the episode, for lack of anything but circumstantial evidence, his view is that the male bonobo suffered fatal injuries.

On another occasion, Hohmann thinks that he came close to seeing infanticide, which is also generally ruled to be beyond the bonobo's behavioral repertoire. A newborn was taken from its mother by another female; Hohmann saw the mother a day later. This female was carrying its baby again, but the baby was dead. "Now it becomes a criminal story," Hohmann said, in a mock-legal tone. "What could have happened? This is all we have, the facts. My story is the unknown female carried the baby but didn't feed it and it died." Hohmann has made only an oblique reference to this incident in print.

These tales of violence do not recast the bonobo as a brute. (Nor does new evidence from Lui Kotal that bonobos hunt and eat other primates.) But such accounts can be placed alongside other challenges to claims of sharp differences between bonobos and chimpanzees. For example, a study published in 2001 in the *American Journal of Primatology* asked, "Are Bonobos Really More Bipedal Than Chimpanzees?" The answer was no.

The bonobo of the modern popular imagination has something of the quality of a prescientific great ape, from the era before live specimens were widely known in Europe. An Englishman of the early eighteenth century would have had no argument with the thought of an upright ape passing silent judgment on mankind and driven by an uncontrolled libido. But during my conversation with Jeroen Stevens, in Belgium, he glanced into the zoo enclosure, where a number of hefty bonobos were daubing excrement on the walls, and said, "These bonobos are from Mars. There are many days when there is no sex. We're running out of adolescents." (As de Waal noted, the oldest bonobo in his San Diego study was about fourteen, which is young adulthood; all but one episode of oral sex there involved juveniles; these bonobos also accounted for almost all of the kissing.)

Craig Stanford, in a 1997 study that questioned various alleged bonobo-chimpanzee dichotomies, wrote, "Female bonobos do not mate more frequently or significantly less cyclically than chimpanzees." He also reported that male chimpanzees in the wild actually copulated more often than male bonobos. De Waal is unimpressed by Stanford's analysis. "He counted only heterosexual sex," he told me. "But if you include all the homosexual sex, then it's actually quite different." When I asked Hohmann about the bonobo sex at Lui Kotal, he said, "It's nothing that really strikes me." Certainly, he and his team observe female g-g rubbing, which is not seen in chimpanzees and needs to be explained. "But does it have anything to do with sex?" Hohmann asked. "Probably not. Of course, they use the genitals, but is it erotic behavior or a greeting gesture that is completely detached from sexual behavior?"

A hug? "A hug can be highly sexual or two leaders meeting at the airport. It's a gesture, nothing else. It depends on the context."

At Lui Kotal, the question of dominance was also less certain than one might think. When I'd spoken to de Waal, he had said, unequivocally, that bonobo societies were dominated by females. But in Hohmann's cautious mind, the question is still undecided.

Data from wild bonobos are still slight, and science still needs to explain the physical superiority of males: why would evolution leave that extra bulk in place, if no use was made of it? Female spotted hyenas dominate male hyenas, but they have the muscle to go with the lifestyle (and, for good measure, penises). "Why hasn't this leveled out in bonobos?" Hohmann asked. "Perhaps sometimes it *is* important" for the males to be stronger. "We haven't seen accounts of bonobos and leopards. We don't know what protective role males can play." Perhaps, Hohmann went on, males exercise power in ways we cannot see: "Do the males step back and say to the females, 'I'm not competing with you, you go ahead and eat'?" The term "male deference" has been used to describe some monkey behavior. De Waal scoffs: "Maybe the bonobo males are chivalrous! We all had a big chuckle about that."

Hohmann mentioned a recent experiment that he had done in the Frankfurt zoo. A colony of bonobos was put on a reduced-calorie diet for the purpose of measuring hormones in their urine at different moments in their fast. It was not a behavioral experiment, but it was hard not to notice the actions of one meek male. "This is a male that in the past has been badly mutilated by the females," Hohmann said. "They bit off fingers and toes, and he really had a hard life." This male had always been shut out at feeding time. Now, as his diet continued, he discovered aggression. "For the first time, he pushed away some low-ranking females," Hohmann said. He successfully fought for food. He became bold and demanding. A single hungry animal is not a scientific sample, but the episode showed that this male's subservience was, if not exactly a personal choice, one of at least two behavioral options.

The media still regularly ask Frans de Waal about bonobos, and he still uses the species as a stick to beat what he scorns as "veneer theory" — the thought that human morality is no more than a veneer of restraint laid over a vicious animal core. Some of his colleagues in primatology admit to impatience with his position — and with the broader bonobo cult that flattens a complex animal into a caricature of Edenic good humor. "Frans has got all the best intentions, in all sorts of ways, but there is this sense in which this polarizing of chimps and bonobos can be taken too far," Richard Wrangham said. Hohmann concurred: "There are certainly some points where we are in agreement; and there are other points where I say, 'No, Frans, you should go to Lomako or Lui Kotal and watch bonobos, and then you'd know better.'" He went on, "Frans enjoyed the luxury of being able to say field work is senseless. When you see wild bonobos, some things that he has emphasized and stretched are much more modest; the sex stuff, for example. But other things are even more spectacular. He hasn't seen meat sharing, he has never seen hunting."

"I think Frans had free rein to say anything he wanted about bonobos for about ten years," Stanford told me. "He's a great scientist, but because he's worked only in captive settings this gives you a blindered view of primates. I think he took a simplistic approach, and, because he published very widely on it and writes very nice popular books, it's become the conventional wisdom. We had this large body of evidence on chimps, then suddenly there were these other animals that were very chimplike physically but seemed to be very different behaviorally. Instead of saying, 'These are variations on a theme,' it became point-counterpoint." He added, "Scientific ideas exist in a marketplace, just as every other product does."

At the long table in the center of the camp, I showed Hohmann the SAVE THE HIPPIE CHIMPS! flier from the Manhattan benefit. He was listening on headphones to Mozart's Requiem; he glanced at the card and put it to one side. Then, despite himself, he laughed and picked it up again, taking off the headphones. "Well," he said.

We were at Lui Kotal for three weeks. "If you stay here, the hours become days, become months," Martin Surbeck said. "It all melts." We had two visitors: a Congolese official who, joined by a guard carrying an AK-47, walked from a town twenty-five miles away to cast an eye over the camp and accept a cash consideration. He stayed for twenty-four hours; every hour his digital wristwatch spoke the time, in French, in a woman's voice — "*Il est deux heures.*"

I saw the bonobos only one other time. I was in the forest with Brigham Whitman, one of the two bearded Americans, when we heard a burst of screeching. In a whisper, Whitman pointed out Dante, a senior male, sitting on a low branch. "He's one of the usual suspects," Whitman said. "Balls hanging out, that's his pose." Whitman ran through Dante's distinguishing characteristics: "He's very old — perhaps thirty — and missing most of his right index finger. His lips are cracked and his face is weathered, but his eyes are vibrant. He has large white nipples. His toes are extremely fat and huge, and his belly hair is redder." He was the oldest male. "Dante just gets his spot and he doesn't move. He just sits and eats."

We followed Dante and a dozen others throughout the afternoon. They climbed down from trees, walked, and climbed back up. Small, nonstinging bees congregated in the space between our eyeballs and the lenses of our binoculars. In the late afternoon, Dante and others climbed the highest trees I had seen in the forest. It was almost dark at the forest floor, but the sun caught the tops of the trees, and Dante, 150 feet up, gazed west, his hair looking as if he'd just taken off Darth Vader's helmet, his expression grave.

In the lobby of the Grand Hotel in Kinshasa, the Easter display was a collection of dazed live rabbits and chicks corralled by a low white wicker fence. At an outdoor bar, the city's diplomatic classes gave each other long-lasting handshakes while their children raced around a deep, square swimming pool. I sat with Gottfried Hohmann; we had hiked out of Lui Kotal together the day before. As we left the half-light of the forest to reach the first golden patch of savanna and the first open sky, it had been hard not to feel evolutionary stirrings, to feel oneself speeding through an "Ascent of Man" illustration, knuckles lifting from the ground.

By the pool, Hohmann talked about a Bavarian childhood collecting lizards and reading Konrad Lorenz. He was glad to be going home. He has none of the fondness for Congo that he once had for India. Still, he will keep returning until retirement. He said that in Germany, when he eats dinner with friends who work on faster-breeding, more conveniently placed animals, "I think, Oh, they live in a different world! People say, 'You're *still* . . . ?' I say, 'Yes. Still.' This big picture of the bonobo is a puzzle, with a few pieces filled, and these big white patches. This is still something that attracts me. This piece fits, this doesn't fit, turning things around, trying to close things."

Because of Hohmann's disdain for premature theories, and his data-collecting earnestness, it had sometimes been possible to forget that he is still driving toward an eventual glimpse of the big picture — and that this picture includes human beings. Humans, chimpanzees, and bonobos share a common ancestor. Was this creature bonobo-like, as Hohmann suspects? Did the ancestral forest environment select for male docility, and did *Homo* and the chimpanzee then both dump that behavior, independently, as they evolved in less bountiful environments? The modern bonobo holds the answer, Hohmann said; in time, its behavior will start to illuminate such characteristics as relationships between men and women, the purpose of aggression, and the costs and benefits of male bonding.

At Lui Kotal, there were no rocks in the sandy earth, and the smallest pebble on a riverbed had the allure of precious metal. It is not a place for fossil hunters; the biological past is revealed only in the present. "What makes humans and nonhuman primates different?" Hohmann said. "To nail this down, you have to know how these nonhuman primates behave. We have to measure what we can see today. We can use this as a reference for the time that has passed. There will be no other way to do this. And this is what puts urgency into it: because there is no doubt that in a hundred years, there won't be great apes in the wild. It would be blind to look away from that. In a hundred years, the forest will be gone. We have to do it now. This forest is the very, very last stronghold. This is all we have."

TODD PITOCK

Science and Islam in Conflict

FROM *Discover*

CAIRO, EGYPT. "There is no conflict between Islam and science,"
Zaghloul El-Naggar declares as we sit in the parlor of his villa in
Maadi, an affluent suburb of Cairo. "Science is inquisition. It's running after the unknown. Islam encourages seeking knowledge. It's
considered an act of worship."

What people call the scientific method, he explains, is really
the Islamic method: "All the wealth of knowledge in the world
has actually emanated from Muslim civilization. The Prophet Mohammed said to seek knowledge from the cradle to the grave.
The very first verse came down: 'Read.' You are required to try to
know something about your creator through meditation, through
analysis, experimentation, and observation."

Author, newspaper columnist, and television personality, El-
Naggar is also a geologist whom many Egyptians, including a number of his fellow scientists, regard as a leading figure in their community. An expert in the somewhat exotic topic of biostratification
— the layering of Earth's crust caused by living organisms — El-
Naggar is a member of the Geological Society of London and publishes papers that circulate internationally. But he is also an Islamic
fundamentalist, a scientist who views the universe through the lens
of the Koran.

Religion is a powerful force throughout the Arab world — but
perhaps nowhere more so than here. The common explanation is
that the Egyptian people, rich and poor alike, turned to God after
everything else failed: the mess of the government's socialist experiment in the 1960s, the downfall of Gamal Abdel Nasser's Arab na-

tionalism, the military debacle of the 1967 war with Israel, poverty, inept government — the list goes on.

I witness firsthand the overlapping strands of history as I navigate the chaos of Cairo, a city crammed with 20 million people, a quarter of Egypt's population. In residential neighborhoods, beautiful old buildings crumble, and the people who live in them pile debris on rooftops because there is no public service to take it away. Downtown, luxury hotels intermingle with casinos, minarets, and even a Pizza Hut. The American University in Cairo is a short distance from Tahrir Square, a wide traffic circle where bruised old vehicles brush pedestrians who make the perilous crossing. At all hours men smoke water pipes in city cafés; any woman in one of these *qawas* would almost certainly be a foreigner. Most Egyptian women wear a veil, and at the five designated times a day when the muezzins call, commanding Muslims to pray, the men come, filling the city's mosques.

The Islamic world looms large in the history of science, and there were long periods when Cairo — in Arabic, El Qahira, meaning "the victorious" — was a leading star in the Arabic universe of learning. Islam is in many ways more tolerant of scientific study than is Christian fundamentalism. It does not, for example, argue that the world is only 6,000 years old. Cloning research that does not involve people is becoming more widely accepted. In recent times, though, knowledge in Egypt has waned. And who is responsible for the decline?

El-Naggar has no doubts. "We are not behind because of Islam," he says. "We are behind because of what the Americans and the British have done to us."

The evil West is a common refrain with El-Naggar, who, paradoxically, often appears in a suit and tie, although he is wearing a pale green *galabiyya* when we meet. He says that he grieves for Western colleagues who spend all their time studying their areas of specialization but neglect their souls; it sets his teeth on edge that the West has "legalized" homosexuality. "You are bringing man far below the level of animals," he laments. "As a scientist, I see the danger coming from the West, not the East."

He hands me three short volumes he has written about the relationship of science and Islam. These include *The Geological Concept of Mountains in the Holy Koran* and *Treasures in the Sunnah: A Scientific*

Approach, parts one and two, along with a translation of the Koran, whose title page he has signed, although his name does not appear as a translator.

In *Treasures in the Sunnah,* El-Naggar interprets holy verses: the hadiths, or sayings of the Prophet, and the sunnah, or customs. There are scientific signs in more than a thousand verses of the Koran, according to El-Naggar, and in many sayings of the Prophet, although these signs often do not speak in a direct scientific way. Instead, the verses give man's mind the room to work until it arrives at certain conclusions. A common device of Islamic science is to cite examples of how the Koran anticipated modern science, intuiting hard facts without modern equipment or technology. In *Treasures of the Sunnah,* El-Naggar quotes scripture: "And each of them [that is, the moon and the sun] floats along in [its own] orbit." "The Messenger of Allah," El-Naggar writes, "talked about all these cosmic facts in such accurate scientific style at a period of time when people thought that Earth was flat and stationary. This is definitely one of the signs, which testifies to the truthfulness of the message of Mohammed."

Elsewhere he notes the Prophet's references to "the seven earths"; El-Naggar claims that geologists say that Earth's crust consists of seven zones. In another passage, the Prophet said that there are 360 joints in the body, and other Islamic researchers claim that medical science backs up the figure. Such knowledge, the thinking goes, could have been given only by God.

Critics are quick to point out that Islamic scientists tend to use each other as sources, creating an illusion that the work has been validated by research. The existence of 360 joints, in fact, is not accepted in medical communities; rather, the number varies from person to person, with an average of 307. These days most geologists divide Earth's crust into fifteen major zones, or tectonic plates.

El-Naggar even sees moral meaning in the earthquake that triggered the 2005 tsunami and washed away nearly a quarter of a million lives. Plate tectonics and global warming be damned: God had expressed his wrath over the sins of the West. Why, then, had God punished Southeast Asia rather than Los Angeles or the coast of Florida? His answer: because the lands that were hit had tolerated the immoral behavior of tourists.

*

The influence and popularity of El-Naggar — as a frequent guest on Arab satellite television, he reaches an audience of millions — does not sit well with Gamal Soltan, a political scientist at Al-Ahram Center for Political and Strategic Studies, a Cairo-based think tank.

"This tendency to use their knowledge of science to 'prove' that the religious interpretations of life are correct is really corrupting," he tells me. Soltan, who got his doctorate at the University of Northern Illinois, works in a small office that's pungent with tobacco smoke; journals and newspapers lie stacked on his desk and floor. "Their methodology is bad," he says. Soltan explains that Islamic scientists start with a conclusion (the Koran says the body has 360 joints) and then work toward proving that conclusion. To reach the necessary answer they will, in this instance, count things that some orthopedists might not call a joint. "They're sure about everything, about how the universe was created, who created it, and they just need to control nature rather than interpret it," Soltan adds. "But the driving force behind any scientific pursuit is that the truth is still out there."

Researchers who don't agree with Islamic thinking "avoid questions or research agendas" that could put them in opposition to authorities — thus steering clear of intellectual debate. In other words, if you are a scientist who is not an Islamic extremist, you simply direct your work toward what is useful. Scientists who contradict the Koran "would have to keep a low profile." When pressed for examples, Soltan does not elaborate.

The emphasis on utility wasn't always present here. The Napoleonic occupation from 1798 to 1801 brought French scientists to Egypt. The arrival of the Europeans alerted Egyptians to how far behind they'd fallen; that shock set in motion a long intellectual awakening. During the 150 years that followed, institutions of higher learning in Cairo gave the city an international reputation for prestigious institutions, and the exchange of scholars went in both directions, with Egyptians going west and Americans and Europeans coming here.

Then came the 1952 coup led by Gamal Abdel Nasser that toppled King Farouk I. Nasser was the first modern leader to position himself as a spokesman for the whole Arab world. His brand of nationalism was meant to unify all Arab people, not just Egyptians, and it set them in opposition to America and Europe. "After

Nasser, Arab nationalism raised suspicions about the West," Soltan says.

In Soltan's view, the twin forces of Islamization and government policy have inadvertently worked together to blunt scientific curiosity. "We are in a period of transition," he says. "I think we are going to be in transition for a long time."

The people and authorities are still grappling with religion's place in Egyptian society, resulting in a situation similar to one in Europe during the time of Copernicus and Galileo, when scientific knowledge was considered threatening to the prevailing religious power structure. For now, the door on freedom of thought has nearly been shut. As Soltan points out, "Cairo University has not received Western professors since the 1950s, and because of the turmoil in the country, many professors who didn't like the regime were excluded from the university."

I walk the campus of Cairo University prior to meeting Waheed Badawy, a chemistry professor who has taught there since 1967. His students, male and female, wander in and out during our talk; the women all wear head covers, highlighting the degree to which religion is particularly strong among the young. He wears a white lab coat, and there are religious verses posted on his laboratory walls and corkboard. Yet Badawy, who specialized in solar energy conversion while working for Siemens in Germany in the 1980s, does not consider himself an "Islamic scientist" like El-Naggar. He is a scientist who happens to be devout, one who sees science and religion as discrete pursuits.

"Islam has no problems with science," he says. "As long as what you do does not harm people, it is permitted. You can study what you want, you can say what you want."

What about, say, evolutionary biology or Darwinism? I ask. (Evolution is taught in Egyptian schools, although it is banned in Saudi Arabia and Sudan.) "If you are asking if Adam came from a monkey, no," Badawy responds. "Man did not come from a monkey. If I am religious, if I agree with Islam, then I have to respect all of the ideas of Islam. And one of these ideas is the creation of the human from Adam and Eve. If I am a scientist, I have to believe that."

But from the point of view of a scientist, is it not just a story? I ask. He tells me that if I were writing an article saying that Adam and Eve is a big lie, it would not be accepted until I could prove it.

Avenida Naperville
Library

Please Return

THE BEST AMERICAN SERIES® 2008

THE BEST AMERICAN Science AND Nature WRITING™

JEROME GROOPMAN, M.D.

EDITOR

TIM FOLGER

SERIES EDITOR

"Nobody can just write what he thinks without proof. But we have real proof that the story of Adam as the first man is true."

"What proof?"

He looks at me with disbelief: "It's written in the Koran."

Tunis, Tunisia. After the hazy congestion of Cairo, the briny sea breeze and open spaces of Tunis are liberating. Anchored on the Mediterranean coast, Tunisia's capital is rimmed by mountainous suburbs with palm trees and gardens trellised with bougainvillea. The town where I am staying is Sidi Bou Said. It has a kind of high-rent antiquity that feels like Italy or the south of France. Indeed, just eighty miles from Sicily, Tunis is physically closer — and culturally closer, too, many people say — to Mediterranean Europe than it is to much of the rest of the Arab world. "They're not really Arabs," my Egyptian translator says en route to the airport. "They're French." He does not mean it as a compliment.

"We have succeeded in keeping extremism and that mentality out of our schools and institutions," says a government official who asks not to be named. "We are an island of ten million people in a sea of Islamists. The extremists want to remove the buffer between religion and everything else, including science. There has to be a buffer between religion and science."

Tunisia, a former French protectorate that became independent in 1956, shares with its Arab neighbors a poor human rights record and a president whose family has been charged with corruption. Freedom House, a nonprofit monitoring group, ranks it 179 out of 195 countries for press freedom. In March 2007 a dissident was sentenced to three and a half years in prison (having already served two years while awaiting trial) for decrying the lack of freedom. Yet unlike the Egyptians, who complain openly about their lack of freedom, the Tunisians I encounter tend to put things in a more optimistic light. One reason for the allegiance to their government is a widely held belief that the alternative to their president, Ben Ali, would be Islamic extremists. Another reason many support the government is that it has been more effective than those of most Arab countries at delivering basic services, including education and health care.

Although officially Muslim, Tunisia maintains the closest thing there is in the Arab world to separation of mosque and state. In public-sector jobs, beards and veils are banned. On the street, you see young women with their hair covered, but it is not unusual for the same women to be wearing tight jeans, making the veil as much fashion accessory as religious garment. School textbooks lack information on different religions and religious beliefs. Islamic science is not a university subject here, as it is in Egypt; Islamology, which looks critically at Islamic extremism, is.

In contrast to the situation in Egypt, where even the most Western-oriented scientist I talked to at some point or other declared himself to be "a good Muslim," in Tunisia the personal religious views of scientists I meet hardly seem relevant. Even so, I am reminded how science, like politics, tends to be local, addressing immediate problems using materials at hand. Sami Sayadi, director of the bioprocesses lab at the Biotechnology Center of Sfax, Tunisia's second-largest city, spent more than a decade figuring out how to turn the waste of olives pressed for oil into clean, renewable energy. Olives have been a major export here since the heyday of Carthage and remain an icon for Arabs everywhere, making Sayadi's achievement sound almost like modern-day alchemy.

Sayadi's thinking is the kind of pragmatism the Tunisian government wants, and in recent years it has come to see science and technology as important tools of national advancement. There were 139 laboratories across different disciplines in 2005, compared with 55 in 1999. The government is actively promoting this growth.

Ninety minutes south of Tunis is the Borj-Cedria Science and Technology Park, a campus that will eventually combine an educational facility, an industrial and R&D center, and a business incubator. The park's completion is still years away, however, and although some buildings and labs are in place, geologists, physicists, and other scientists laboring here work with equipment that in the West wouldn't pass muster in many high schools. They pursue projects for the love of science.

The situation may soon change. In its hunger for patents and profits, the Tunisian government is giving out four-year contracts to labs whose work has industrial applications. Senior researchers at Borj-Cedria currently make about $1,100 a month (a livable but modest wage here), but the new program would give anyone who earns a patent a 50 percent stake in royalties.

Still, Tunisia's support of science has clear limits: projects whose aim is solely to advance knowledge get no support. "Everyone would like to do [basic] research," says Taieb Hadhri, the minister of scientific research, technology, and competency development, who has held the cabinet-level post since the department was created in 2004. "I'm a mathematician by training, and I would also like to do [basic] research. But that will have to come later. We have more pressing needs now."

And the push toward advancement here is not entirely free from the pull of tradition, as I learn when I visit Habiba Bouhamed Chaabouni, a medical geneticist who splits her time between research and teaching at the Medical Faculty of the University of Tunis and seeing patients at the Charles Nicolle Hospital, also in the capital. In 2006 she won a L'Oréal-UNESCO Women in Science Award, a $100,000 prize given to five women, each representing one of the continents, for her work analyzing and preventing hereditary disorders. When she greets me in her office, she is wearing a white lab coat. Test tubes clink as they spin in a centrifuge to separate strands of a patient's DNA that Chaabouni will examine later.

Chaabouni recalls the early days of her career, in the mid-1970s, when she saw children afflicted with disfiguring diseases. "It was very sad," she says. "I met families with two, three, four affected siblings. I wanted to do something about it, to know how to prevent it." There was no facility for genetic research at that time, and for two decades, she lobbied government officials hard for it. "We wanted better conditions and facilities. They also saw we were publishing in international [peer-reviewed] journals. I think the policymakers finally understood the value of developing research."

The Tunisian medical genetics community, which includes about a hundred doctors and technicians, now publishes more than any other Arab country. "We looked on PubMed, and we're ahead of Egypt," Chaabouni says, beaming. "Not by a lot, but remember, we're one-tenth the size."

Over the last thirty years, Chaabouni has also seen how people who once resisted her message have begun listening. Once, genetic counseling or even coming in for certain treatments almost amounted to a social taboo; now, it is becoming more accepted, and things that were once simply ignored or not spoken of — such as autism in children, which is being identified more commonly — are more often out in the open.

For all that, Chaabouni still sees how her advice sometimes clashes with her patients' beliefs. Like many Arab and Muslim countries, Tunisia has a high incidence of congenital diseases, including adrenal and blood disorders, that Chaabouni has traced to consanguinity.

"It's a custom here, and in the rest of the Arab world, to marry cousins, even first cousins," she tells me, though the practice is becoming less common. "Of course, that means they share a lot of genes from common sets of grandparents."

In other fields, pure research does not get support; in medical genetics, even practically applicable knowledge can spark conflicts with Islamic culture. "Taking a blood sample to study abnormalities is not a problem," Chaabouni says. "That's just investigation. The problem is when you take the results of research into the clinic and try to give genetic counseling to patients. Then you have people who won't accept the idea that they have to stop having children or that they shouldn't marry their cousin."

Today prenatal screening and genetic testing are more widely accepted, and when it's necessary to save the mother's life, doctors terminate pregnancies. Islamic law permits abortion in cases of medical necessity (where the mother's life is in jeopardy) until 120 days in utero, at which point it regards the fetus as "ensouled" and abortion becomes homicide. For Chaabouni, the challenge is mainly one of communication. "They look for arguments why you may be wrong," she says. "They go to other doctors. In the end, they usually follow our advice, but it's hard because you're giving them bad news that may also go against what they believe."

Mohammed Haddad, an Islamology specialist at the Université de la Manouba in Tunis, points out the many little assaults that can turn people's minds against scientific advances. For example, a sheik recently declared that he'd found a cure for AIDS — spelled out in the Koran. "He was from Yemen, but they reach us by satellite, and it's all a big business," Haddad says. "People listen, and it's a problem. In this situation, many will die."

Amman, Jordan. "The Koran says, 'Read,' but it does not even say 'Read the Koran.' Just 'Read,'" says Prince El Hassan bin Talal, who greets me at the Royal Scientific Society, Jordan's largest research institution — one that he helped establish in 1970. Hassan was heir to the throne until his brother, King Hussein, bypassed him in favor of Abdullah, Hussein's own son. The sixty-year-old prince, who speaks classical Arabic and Oxford English and has studied biblical Hebrew, can tick off a whole list of things that are wrong with Jordan, from Western governments and nongovernmental organizations that come proposing solutions without having identified the causes of problems, to a culture that does not value reading. He is bookish himself; during our forty-minute-plus interview, he refers to Kierkegaard, Karen Armstrong's *A History of God,* and *What Price Tolerance,* a 1939 book by his wife's relative Syud Hossain.

He is also candid, calling suicide bombers "social rejects" and questioning the validity of those who would take the Muslim world back to the times of the Prophet Mohammed. "Are we talking Islam or *Islamism?*" he asks, pointing out the difference between the religion and those extremists who use the religion to advance their own agendas. "The danger [posed by Islamists] is not only to Christians but also to Islam itself. The real problem is not the Arab-Israel issue but the rise of Islamism."

Science, rather than religion, is the way to ensure a country's future, Prince Hassan believes, and he has made supporting scientific achievement a personal mission for almost forty years. He envisions projects that would promote regional partnerships, even with Israel — an idea that, despite official peace between the countries, remains controversial.

He notes that some important science initiatives are under way. One of the Royal Scientific Society's pursuits is the Trans-Mediterranean Renewable Energy Cooperation, or TREC, a multinational effort that would use wind, water, geothermal, and solar resources to provide renewable energy from Oman to Iceland. If successful, the endeavor would take decades to be realized. Like Moses standing on Mount Nebo (in fact, the site of the Exodus story lies just about twenty minutes outside Amman), the sixty-year-old Hassan knows that he is not likely to see this technological promised land himself.

"Vision," he says, "is not an individual thing. It's a collaboration."

"The biggest disaster in the region, I am sorry to say, is the loss of brainpower," admits Hassan. The emigration of trained academics plagues the entire Arab world, and half of those students who graduate from foreign universities never return to the Arab states. "A large percentage of [America's] NASA staff are of Middle Eastern origin," Hassan notes.

In some ways, the brain drain in Jordan is more obvious than in Egypt because resources here are stretched to the breaking point. Conservative estimates put the number of Iraqi refugees living in Jordan at 700,000 — an enormous burden considering that Jordan has just 6 million citizens. To put that figure in perspective, imagine the United States adding 35 million people in a period of four years.

The population influx has triggered inflation, soaring rents and property prices, and urban sprawl. Like Egypt, Tunisia, and Syria (and like Israel, for that matter), Jordan lacks significant natural resources; the country has little oil or fresh water. In fact, since most of the water from the Jordan River's tributaries has been diverted and no longer flows to the Dead Sea, even the Dead Sea is dying. There are plans for resuscitating it, but they will require a delicate process of regional cooperation, including the Israelis and the Palestinians, and most likely Western aid.

Jordan also lacks financial resources, unlike the oil-rich Gulf states, which can afford to treat knowledge and expertise as an accessible commodity that can be imported as needed. Furthermore, the perception of danger — terrorists bombed three hotels in Amman in 2005, and Al Qaeda has admitted to killing an American diplomat — has all but shut the valve on Jordanian tourism and the considerable revenue it used to bring.

Jordan, the quip goes, is caught between Iraq and a hard place. For now, it embodies many of the issues that the Arab Human Development Report blamed for the region's intellectual malaise, among them lack of freedom and dysfunctional, authoritarian governments whose security services have too much say; the triumph of who-you-know advancement over merit-based promotion; and poor communication between researchers within the region. Educational opportunities are limited, especially for girls and women. All of this means that if you are a talented scientist, there is a good chance you'll leave.

"Science needs stability, democracy, freedom of expression," says Senator Adnan Badran, who has a Ph.D. in molecular biology from Michigan State University, as we drink Turkish coffee at his office. "You must have an environment that's conducive to free thinking, to inquiry. If you don't, you'll never be able to release the mind's potential. It's a very bleak story, a very disappointing story, about the state of science and technology in the Arab region."

He blames a tradition that began with the Ottomans in the 1500s: lowering educational standards and promoting dogma. "We were open. Islam was open, a strong belief with dialogue. It was tolerant, mixing with other civilizations. Then we shifted to being dogmatic. Once you're dogmatic, you are boxed in," he says. "If you step outside the box, you're marginalized — and then you're out. So you go west."

That's what Badran did, spending twenty years in France and the United States, where he earned four patents doing research for the United Fruit Company. His work, which focused on retarding ripeness in bananas, has had huge economic impact — billions of dollars, potentially — because it allows the company to ship its crops around the world without spoiling.

Even so, Badran returned home to Jordan, where he took up academic positions, including the presidency of Philadelphia University in Amman. In 1987 he was made the first secretary general of Jordan's Higher Council for Science and Technology and was later appointed to the senate by the king of Jordan, Abdullah II. Then, early in 2005, the king appointed Badran prime minister, the first scientist to hold that position. The king, who was educated at the Royal Military Academy Sandhurst and at Oxford in the United Kingdom and also attended Georgetown University in Washington, D.C., appreciated Badran's position on the need for Arab glasnost.

"I wanted to destroy every vested interest, to get rid of cronyism, to build accountability and transparency by freeing the press," Badran says. The circumstances of Badran's term were difficult, however. "He was an excellent academic and scientist," a journalist tells me, "but an ineffective politician."

Any chance for Badran to advance his agenda went up with the smoke in November 2005 when suicide bombers targeted three Amman hotels. As the government shifted its focus from internal reform to security, Badran was a casualty of change. The prime minister here serves at the discretion of the king — and also, many people say, by tacit approval of Jordan's security services. In less than a year, Badran was ousted (his thinking was considered to be too idealistic for that time) and returned to his seat in the senate.

After leaving Badran, I get a primer on Jordan's most dynamic and hopeful scientific collaboration. I speak with the physicist Hamed Tarawneh at his cramped, dingy temporary office at UNESCO's headquarters in Amman. Tarawneh, a tall, broad-shouldered chain smoker with a disarming smile, left years ago to get his Ph.D. in Sweden and returned to Jordan just a few months prior to our meeting. He is in the process of assembling a staff of engineers and technicians for SESAME (Synchrotron-light for Experimental Science and Applications in the Middle East), an international laboratory organized around a machine that has wide applications in physics, biology, medicine, and archaeology. Only a handful of these versatile light generators exist, and this is the first in the Muslim world.

Jordan was selected as the site for SESAME after King Abdullah II donated land and ponied up $10 million for the facility that would house the synchrotron. The project is modeled on CERN, the Swiss high-energy physics lab formed after World War II to restore Europe's tradition of scientific learning. When SESAME becomes fully operational in 2009 — the facility at Al-Balqa Applied University near Amman was to be complete in June 2007 — researchers will rotate through doing their work in weeks-long sessions. Like its European model, SESAME was conceived in part to motivate the region's best and brightest to stay or even to return from abroad; the laboratory should also create excitement and opportunity that will attract young students to science.

Tarawneh hopes SESAME will become a knowledge hub for the member states that pay annual dues, a group that now includes Bahrain, Egypt, Pakistan, Turkey, the Palestinian Authority — and Israel, the one country in the region that has a knowledge-based society but has been excluded from almost every other endeavor. "We are scientists," Tarawneh says. "We don't care about politics. So now we have a chance to discuss science here and work for the greater good of knowledge. It's a very good start. It's a cosmopolitan environment, which is what we've been lacking. Now we'll all know each other as scientists, as people."

I ask about the legions of scientists who have left Jordan, who regard it as a lost cause.

"Would I earn more if I went to Berkeley?" Tarawneh asks. "Yes, of course. But I am from here. I am an Arab. I am a Muslim. This is where I want to be. And why can't we build something here that's ours? In five years, others will see it's useful, and it will become a world effort and create a culture of scientific inquiry here. Science is the way to break barriers. It's about development and advancing people's interests."

Tarawneh's enthusiasm makes SESAME's success seem inevitable, but the king's support and the international character of the project make it seem like much more than an individual triumph. It is precisely the kind of regional partnership that people like Prince Hassan say is the real road map to peace and prosperity in the Islamic world. As both machine and metaphor, a high-powered generator that shines light on scientific inquiry may be the answer to everyone's prayers.

DAVID QUAMMEN

Deadly Contact

FROM *National Geographic*

IN SEPTEMBER 1994, a violent disease erupted among race-horses in a suburb of Brisbane, Australia. The place, called Hendra, was a quiet old neighborhood filled with racecourses, stables, newsstands that sell tip sheets, corner cafés with names like the Feed Bin, and racing people. The first victim was a pregnant mare named Drama Series, who started showing symptoms in an outlying pasture and was brought back to her trainer's stable for doctoring, where she only got worse. Three people worked to save her — the trainer himself, his stable foreman, and a veterinarian. Within two days Drama Series died, leaving the cause of her trouble uncertain. Had she been bitten by a snake? Had she eaten some poisonous weeds out in that scrubby, derelict meadow? Those hypotheses were eliminated two weeks later, when most of her stablemates fell ill. This wasn't snakebite or toxic fodder. It was something contagious.

The other horses suffered fever, respiratory distress, facial swelling, and clumsiness; in some, bloody froth came from the nostrils and mouth. Despite heroic efforts by the veterinarian, twelve more animals died within days. Meanwhile the trainer himself got sick; so did the stable foreman. The vet, who was following cautionary procedures but working amid the same mad circumstances, stayed healthy. After a few days in a hospital, the trainer died. His kidneys had failed and he couldn't breathe. The stable foreman, a big-hearted man named Ray Unwin, who had merely gone home to endure his fever in private, survived. He and the veterinarian told me their stories when I found them in Hendra last year. Ray Unwin is a middle-aged working bloke with a sandy red ponytail and a weary sadness in his eyes, who professed that he wasn't a "whinger" (complainer) but said his health has been "crook" (not right) since it happened.

Laboratory analysis revealed that the horses and men were infected by a previously unknown virus. At first the lab people called it equine morbillivirus, a horse virus closely related to measles. Later, as its uniqueness became better appreciated, the virus was renamed after the place itself: Hendra. The veterinarian, a tall, gentle fellow named Peter Reid, told me that "the speed with which it went through those horses was unbelievable." At the height of the crisis, seven animals succumbed to ugly deaths or required euthanasia within just twelve hours. One of them died thrashing and gasping so desperately that Reid couldn't get close enough to give it the merciful needle. "I'd never seen a virus do anything like that before," he said. A man of understatement, he recalled it as "a pretty traumatic time."

Identifying the new virus was only step one in solving the immediate mystery of Hendra, let alone understanding the case in a wider context. Step two involved tracking that virus to its hiding place. Where did the thing exist when it wasn't killing horses and people? Step three entailed asking a further cluster of questions: how did it emerge from its secret refuge, and why here, and why now?

After our first conversation, Peter Reid drove me out to the site where Drama Series took sick. Tract houses on prim lanes have been built over the original pasture. Not much of the old landscape remains. But toward the end of one street is a circle, called Calliope Circuit, in the middle of which stands a single mature tree, a native fig, beneath which the mare would have found shelter from eastern Australia's fierce subtropical sun.

"That's it," Reid said. "That's the bloody tree." That's where the bats gathered, he meant.

Infectious disease is all around us. Infectious disease is a kind of natural mortar binding one creature to another, one species to another, within the elaborate edifices we call ecosystems. It's one of the basic processes that ecologists study, including also predation, competition, and photosynthesis. Predators are relatively big beasts that eat their prey from outside. Pathogens (disease-causing agents, such as viruses) are relatively small beasts that eat their prey from within. Although infectious disease can seem grisly and dreadful, under ordinary conditions it's every bit as natural as what lions do to wildebeests, zebras, and gazelles.

But conditions aren't always ordinary.

Just as predators have their accustomed prey species, their favored targets, so do pathogens. And just as a lion might occasionally depart from its normal behavior — to kill a cow instead of a wildebeest, a human instead of a zebra — so can a pathogen shift to a new target. Accidents happen. Aberrations occur. Circumstances change, and with them opportunities and exigencies also change. When a pathogen leaps from some nonhuman animal into a person and succeeds in making trouble there, the result is what's known as a zoonosis.

The word "zoonosis" is unfamiliar to most people. But it helps clarify the biological reality behind the scary headlines about bird flu, SARS, other forms of nasty new diseases, and the threat of a coming pandemic. It says something essential about the origin of HIV. It's a word of the future, destined for heavy use in the twenty-first century.

Ebola is a zoonosis. So is bubonic plague. So are yellow fever, monkeypox, bovine tuberculosis, Lyme disease, West Nile fever, Marburg, many strains of influenza, rabies, hantavirus pulmonary syndrome, and a strange new affliction called Nipah, which kills pigs and pig farmers in Malaysia. Each of them reflects the action of a pathogen that can cross to people from other species. This form of interspecies leap is common, not rare; about 60 percent of all human infectious diseases currently known are shared between animals and humans. Some of those — notably rabies — are widespread and famously lethal, still killing humans by the thousands despite centuries of efforts to cope with their effects, concerted international attempts to eradicate or control them, and a clear scientific understanding of how they work. Others are new and inexplicably sporadic, claiming a few victims (as Hendra did) or a few hundred in this place or that, and then disappearing for years.

Smallpox, to take one counterexample, is not a zoonosis. It's caused by a virus that infects *Homo sapiens* and, in very exceptional cases, certain nonhuman primates, but not horses or rats or other species. That helps explain why the World Health Organization's global campaign to eradicate the disease was, as of 1979, successful. Smallpox could be eradicated because its virus, lacking the ability to reside virtually anywhere other than in humans, couldn't hide. Zoonotic pathogens can hide.

Monkeypox, though closely related to smallpox, differs in two crucial ways: its propensity to afflict monkeys as well as humans and the ability of its virus to exist in still other species, some of which are so far unidentified. Yellow fever, also infectious to both monkeys and humans, and caused by a virus that hides in several species of mosquito, will probably never be eradicated. The Lyme disease perpetrator, a type of bacterium, hides effectively in white-footed mice and other small mammals. These pathogens aren't consciously hiding, of course. For their purposes, such behavior merely constitutes a strategy of indirect transmission or inconspicuous survival.

The least conspicuous strategy of all is to lurk within what's called a reservoir host, a species that carries the pathogen while suffering little or no symptomatic illness. When a disease seems to disappear between outbreaks (again, as Hendra did after the 1994 carnage), its causal pathogen may indeed have died out, at least from the region — but then again, maybe not. Maybe it is still lingering nearby, all around, within some reservoir host. A rodent? A bird? A butterfly? Possibly a bat? To reside undetected within a reservoir host is probably easiest wherever biological diversity is high and the ecosystem is relatively undisturbed. The converse is also true: ecological disturbance causes diseases to emerge. Shake a tree, and things fall out.

Some months after the deaths in Australia, a scientific sleuth named Hume Field started looking for Hendra's reservoir host. Field was a veterinarian who, having practiced privately for years, had decided to pursue a doctorate in veterinary epidemiology. The search for the reservoir became his dissertation project. He gathered blood samples from sixteen different species, a whole menagerie of suspects, including marsupials, birds, rodents, amphibians, and insects. He sent the samples to a laboratory for screening, which yielded no evidence whatsoever of Hendra.

Then he took blood from *Pteropus alecto,* a species of fruit bat as big as a crow and commonly known as the black flying fox. Bingo: the lab team found molecular traces left by Hendra virus. Further sampling produced similar evidence from three other species of flying foxes, all native to the forests of Queensland (the state encompassing Brisbane) and other wooded regions of Australia. Field and his collaborators had established that bats were the reservoir. Detecting molecular traces is less definitive than finding particles of live virus, but within one female bat they did find that form of evidence also.

The lab work suggested that Hendra was an old virus, having probably existed within its reservoir host for thousands of years. Despite its age, it had never before — so far as historical records and human memory could say, anyway — caused disease in humans. What accounted for its emergence in 1994? Well, bad luck for Drama Series and those who knew her. Bats came to eat the figs in that solitary tree, and the poor mare, seeking shade, grazing too carelessly, evidently swallowed not just grass but also something of what they dropped, such as fruit pulp, feces, urine, afterbirth, and virus.

But there had to be a broader answer, too. Why did Hendra emerge in 1994, not decades or centuries earlier? Something was different. Some sort of change, or combination of changes, must have caused the virus to be transferred from its reservoir host to other species.

The fancy name for such transfer is spillover. Maybe the virus needed horses (which reached Australia only with European colonists), as distinct from kangaroos (which have been eating grass beneath Australian fig trees for millennia), to mediate its spillover from the reservoir. Maybe bats, figs, horses, and humans had simply never been brought so close together. Hume Field is currently a research scientist at the Animal Research Institute of Queensland's Department of Primary Industries in Brisbane. When I spoke with him at his office there, he raised the issue of "what might be happening now that hasn't happened before." Part of the answer is that human destruction of eucalyptus forests has disrupted the customary feeding and roosting habits of some flying foxes, forcing them toward shady suburbs, orchards, botanical gardens, city parks, and closer proximity to people.

But proximity is one thing; spilling virus into horses is another. "How does transmission occur?" Field wondered aloud, at the end of our long conversation. "Well, we still don't know."

*

Nearly all zoonotic diseases result from infection by one of six kinds of pathogens: viruses, bacteria, protozoans, prions, fungi, and worms. Mad cow disease is caused by a prion, a weirdly folded protein molecule that triggers weird folding in other molecules, like the infectious form of water, ice-nine, in Kurt Vonnegut's great early novel *Cat's Cradle*. Sleeping sickness is a protozoan infection, carried by tsetse flies between wild and domestic mammals and people in the landscapes of sub-Saharan Africa. Anthrax is a bacterium that can live dormant in soil for years and then, when scuffed out, infect humans by way of cattle. Toxocariasis is a mild zoonosis caused by roundworms; you can get it from your dog. But fortunately, like your dog, you can be wormed.

Viruses are the most problematic. They evolve quickly, they are unaffected by antibiotics, they can be elusive, they can be versatile, they can inflict extremely high rates of mortality, and they are fiendishly simple, at least relative to other living or quasi-living creatures. Hanta, SARS, monkeypox, rabies, Ebola, West Nile, Machupo, dengue, yellow fever, Junin, Nipah, Hendra, influenza, and HIV are all viruses. The full list is much longer. There is a thing known by the vivid name simian foamy virus (SFV) that infects monkeys and humans in Asia by way of venues (such as Buddhist and Hindu temples) where people and half-tame macaques come into close contact. Some of the people visiting those temples, feeding handouts to those macaques, exposing themselves to SFV, are international tourists. "Viruses have no locomotion," according to the eminent virologist Stephen S. Morse, "yet many of them have traveled around the world." They can't run, they can't walk, they can't swim, they can't crawl. They ride.

About the same time as the Hendra outbreak near Brisbane, another spillover occurred, this one in central Africa. Along the upper Ivindo River in northeastern Gabon, near the border with the Republic of the Congo, lies a small village called Mayibout II. In early February 1996, eighteen people there became suddenly sick after they participated in the butchering and eating of a chimpanzee. Their symptoms included fever, headache, vomiting, bleeding in the eyes, bleeding from the gums, hiccupping, and bloody diarrhea. All eighteen were evacuated downriver to a regional hospital, where four soon died. The bodies were returned to Mayibout II and buried with no special precautions; a fifth victim escaped from the hospital, went back to the village, and died there. Secondary cases occurred among people infected by loved ones or friends or in handling the dead bodies. Eventually thirty-one people got sick, of whom twenty-one died — a mortality rate of 68 percent.

Those facts and numbers were collected by a team of medical researchers, some Gabonese, some French, who reached Mayibout II during the outbreak. Among them was a Frenchman named Eric M. Leroy, based at the Centre International de Recherches Médicales de Franceville (CIRMF), in Franceville, Gabon. Leroy and his colleagues identified the disease as Ebola hemorrhagic fever and deduced that the butchered chimpanzee had been infected with Ebola virus. Their investigation also revealed that the chimp hadn't been killed by village hunters; it had been found dead in the forest and scavenged.

Four years later, I sat at a campfire near the upper Ivindo River with a group of local men working as forest crew for a long overland trek. (See "The Green Abyss: Megatransect, Part II," *National Geographic*, March 2001.) The men, mostly Bantu, had been walking for weeks before I joined them on the march. Their job involved carrying heavy bags through the jungle and building a new camp every night for the Wildlife Conservation Society biologist J. Michael Fay, whose extraordinary grit and sense of mission drove the enterprise forward. This particular day had been a relatively easy one — no swamps crossed, no charging elephants — which allowed for a relaxed, confiding atmosphere at the evening fire. I learned that two of the men, Thony M'Both and Sophiano Etouck, had been present in Mayibout II when Ebola struck the village. M'Both, slim in build, older and more voluble than the others, was willing to talk about it. He spoke in French, while Etouck, a shy man with wide shoulders, an earnest scowl, and a goatee, sat silent. Etouck's own family had been devastated by the disease. He had held one of his dying nieces in his arms while an IV drip in her wrist became clogged, swelled her hand, and exploded, covering him with her blood. Yet Etouck himself never got sick. Nor did I, said M'Both. The cause of the illnesses was a matter of confusion and fearful rumor. M'Both suspected that French soldiers, visiting nearby, had killed the chimpanzee with some sort of chemical weapon and carelessly left it to poison unsuspecting people. But whatever the cause, whatever the contaminant, his fellow villagers had learned their lesson. To this day, he said, no one in Mayibout II eats chimpanzee.

Amid the chaos and sorrow of the outbreak, M'Both told me, he and Etouck had seen something bizarre: thirteen gorillas, all dead, lying in the forest. That image, of thirteen gorilla carcasses strewn on the leaf litter, is lurid but plausible. Subsequent research has confirmed that gorillas are susceptible to Ebola. Being social creatures, they could easily pass the infection among group members by mutual grooming, infant care, or trying to rouse their sick or their dead.

In the years since 1996, other outbreaks of Ebola have struck both people and great apes (chimps as well as gorillas) within the region surrounding Mayibout II. One area hit hard lies along the Mambili River, just over the border in northwestern Congo, another zone of dense forest encompassing several villages, a national park, and a gorilla sanctuary known as Lossi. Mike Fay and I walked through that area also, in March 2000, during one of my earlier stints with his expedition. At the time, gorillas were abundant within the Mambili drainage. But in 2002 a team of researchers at Lossi began finding gorilla carcasses, some of which tested positive for Ebola. Within a few months, 91 percent of the individual gorillas they'd been studying (130 of 143 animals) had vanished, and most of those were presumably dead. Extrapolating from confirmed deaths and disappearances to the overall toll throughout their study area, the researchers published a paper in *Science* under the headline EBOLA OUTBREAK KILLED 5,000 GORILLAS.

In the autumn of 2006 I returned to the Mambili River with a team led by William B. (Billy) Karesh, the director of the Wildlife Conservation Society's Field Veterinary Program and an authority on zoonotic diseases. Karesh's goal was to tranquilize a few surviving gorillas, take blood samples, and see whether those animals showed exposure to Ebola. Along with an expert tracker named Prosper Balo, plus other veterinarians and guides, we spent eight days searching the forest. Prosper Balo had worked at Lossi. With his guidance, we staked out a *bai* (a natural clearing) full of succulent vegetation, previously known for the dozens of gorillas that came there daily to eat and relax. Billy Karesh himself had visited

the same area in 2000, before Ebola struck, to gather baseline data
on gorilla health. "Every day," he told me, "every bai had at least a
family group." He'd been successful on that trip — the only person
ever to tranquilize-dart lowland gorillas. This time things were different. So far as we could see, there scarcely *were* any survivors. We
caught glimpses of just two gorillas. The others had either dispersed to parts unknown or they were . . . dead? Anyway, once gorillas had been abundant hereabouts, and now they were gone.

The virus seemed to be gone, too. But we knew it was only hiding.

Hiding where? For a decade, the identity of Ebola's reservoir host
was one of the darkest mysteries in the world of disease science.
Several sets of researchers were trying to solve it. Then, in 2005,
Eric Leroy and some colleagues announced in the journal *Nature,*
"We find evidence of asymptomatic infection by Ebola virus in
three species of fruit bat, indicating that these animals may be
acting as a reservoir for this deadly virus." Leroy's group hadn't
captured any live virus, but they had established — with positive results from several kinds of molecular tests — that Ebola had passed
through at least a few of the bats examined.

Leroy himself wants stronger evidence. "We continue to catch
bats — to try to isolate virus from their organs," he said late last
year when I visited him in Franceville. Identifying the reservoir
host with certitude, though, would still leave other questions unanswered.

For instance, how does Ebola emerge from that reservoir? "We
don't know if there's direct transmission from bats to humans,"
Leroy said. "We only know there is direct transmission from dead
great apes to humans." And how has the virus evolved, producing
four distinct strains? Why is the Ebola-Zaire strain, the one found
in Gabon and Congo, so highly lethal (about 80 percent mortality)
to people? What is its natural life cycle? What's the spillover mechanism into gorillas and chimps? How does the virus affect the human immune system? How does it find its way into humans at all?
Ebola is difficult to study, Leroy explained, because of the character of the disease. It strikes rarely, it progresses quickly, it kills or it
doesn't kill within just a few days, it affects relatively few people in
each outbreak, and those people generally live in remote, forested

areas, far from research hospitals or medical institutes; then it exhausts itself locally or is successfully stanched and disappears back into the forest, like a hit-and-run force of guerrilla warriors. "There is nothing to do," Leroy said, with the perplexity of a patient man. He meant, nothing to do except keep trying, keep working in the lab, keep responding to outbreaks when they occur. No one can predict where Ebola might next appear. "The virus seems to decide for itself."

Hendra and Ebola are part of a much larger pattern: the recent emergence of new zoonotic diseases, variously lethal and horrific, more than a few of which seem to be associated with bats. Another part of the pattern is human-caused disruption of wild landscape. Nipah came next.

In September 1998, a pork seller in peninsular Malaysia checked into a hospital with some sort of brain inflammation and died. Around the same time, a number of pig-farm workers came down with similar symptoms, a bad fever leading to coma; several of them also died. Pigs in the area meanwhile suffered an illness of their own (or what seemed their own), coughing and wheezing, keeling over dead. The pig disease was taken to be classical swine fever. The human deaths were attributed to Japanese encephalitis. But within a few months, scientists showed that both the pigs and the people had been infected with the same virus, a new one, first isolated from a patient whose home village was called Sungai Nipah. The virus was highly contagious from pig to pig but not from person to person. It spread elsewhere in Malaysia, and even to Singapore, with shipments of live pigs, infecting people who came in contact with the sick animals or their meat. Within seven months, the outbreak had caused 265 human cases, 105 human deaths, and led to the culling of 1.1 million pigs.

The molecular profile of this new virus suggested a close kinship with Hendra. That provided a clue. Not long afterward, researchers found Nipah living sedately in a reservoir host: *Pteropus hypomelanus,* another species of fruit bat. They also noted that fruit bats, deprived of habitat elsewhere, had been congregating in orchards near the pig farms.

And then there was SARS. It came out of southeastern China in early 2003, spreading readily from person to person, traveling as fast as airplanes, killing 774 humans in nine countries and scaring people all over the world. A quick bit of research pointed suspicion at the masked palm civet, a medium-size mammal often sold in Chinese markets for its meat, as the reservoir of SARS. That suspicion was discounted, though, after experiments showed that masked palm civets themselves suffer symptomatic SARS. Then a group of scientists led by Wendong Li, of the Chinese Academy of Sciences, announced that they had found reservoirs hosting a virus very similar to the one that caused the SARS outbreak: horseshoe bats of the genus *Rhinolophus*.

There's more. Australian bat lyssavirus, a newly identified virus closely related to rabies, has killed at least two people with rabies-like symptoms after the victims were bitten by bats. Menangle and Tioman are also bat-carried viruses, of the same family as Hendra, that scientists are watching carefully. Rabies itself and rabies-like viruses, found in bat reservoirs around the world, are still probably the most lethal of all viral pathogens if untreated — with nearly 100 percent mortality among humans. In northern Peru in the autumn of 2006, eleven children from native communities along the Amazon headwaters died from rabies contracted when they were nipped by vampire bats.

At this point, you're entitled to ask, Damn, what is it about bats?

I asked that myself in conversation with Charles Rupprecht, a virologist and veterinarian who leads the rabies section at the Centers for Disease Control and Prevention in Atlanta. Rupprecht recited a list of factors that make this order of mammals, the Chiroptera, ideal candidates to host a variety of dangerous viruses. Some bats roost in huge colonies, snuggled intimately together; they give birth to only a few young and therefore nurture those young dotingly; they have long life spans, especially for small mammals; they are old, too, in evolutionary terms; they encompass a great diversity of species, roughly 20 percent of all mammals; they fly, and therefore they get around the world nicely, finding places and ways to sustain themselves on nearly every landmass except Antarctica. Add to those traits the fact that, being nocturnal and airborne, they're hard to study. "Bats really are the undiscovered country," Rupprecht said. His point — the point of a rabies biologist who happens to like bats — is that they aren't sinister, and if they seem to harbor an undue variety of nightmare diseases, it's probably because they are so various and so poorly known.

Another informed view came from Xavier Pourrut, a research veterinarian based at CIRMF in Gabon. His job involves capturing and taking blood samples from bats near Ebola outbreaks so that Eric Leroy can study their serum for evidence of the virus. "Bats represent an ancient lineage of mammals," Pourrut told me, and like Rupprecht he sees them with a biologist's appreciation. The thing to remember, he said, is that their flight powers give them great range of access, not just horizontally to places around the world, but vertically within the forest. That potentially puts them in contact not just with the fruits or the insects on which they feed and the treetops from which they dangle but also with an inordinate number of other species, from the canopy to the ground, including rodents, monkeys, carnivores, birds, snakes, chimpanzees, gorillas, and people.

Contact is crucial. Close contact between two species represents an opportunity for a pathogen to expand its horizons and possibilities. The pathogen may be well adapted to its quiet, secure life within a reservoir host; spilling over into a new species presents a chance, at some risk, of vastly increasing its abundance and its geographic reach. The risk is that by killing the new host too quickly, before getting itself transmitted onward, the pathogen will come to a dead end. But evolutionary theory suggests that some pathogens on some occasions will accept that risk in exchange for the chance of a big payoff. Long-term survival is only one form of evolutionary success. Gross abundance and broad distribution is another.

Think of tortoises and rats. Tortoises tend to live by a conservative strategy, remaining within their preferred habitat and reproducing slowly. Rats tend to be opportunists, fanning out, traveling across land and sea as stowaways, arriving in new places, and reproducing fast. Similarly, pathogens may differ in their degree of adventurousness. Spillover from a reservoir host isn't necessarily an accident, always leading to the dead end. It may be a strategy leading to evolutionary success. Simian immunodeficiency virus (SIV) achieved that sort of success when it spilled over from one subspecies of chimpanzee into humans, probably in west-central Africa, and became HIV-1.

Close contact between humans and other species can occur in various ways: through the killing and eating of wild animals (as in Mayibout II), through caregiving to domestic animals (as in

Hendra), through fondling of pets (as with monkeypox, brought into the American pet trade by way of imported African rodents), through taming enticements (feeding bananas to the monkeys at a Balinese temple), through intensive animal husbandry combined with habitat destruction (as on Malaysian pig farms), and through any other sort of disruptive penetration of humans into wild landscape — of which, needless to say, there's plenty happening around the world. Once the contact has occurred and the pathogen has crossed over, two other factors contribute to the possibility of cataclysmic consequences: the sheer abundance of humans on Earth, all available for infection, and the speed of our travel from one place to another. When a bad new disease catches hold, one that manages to be transmissible from person to person by a handshake, a kiss, or a sneeze, it might easily circle the world and kill millions of people before medical science can find a way to control it.

But our safety, our health, isn't the only issue. Another thing worth remembering is that disease can go both ways: from humans to other species as well as from them to us. Measles, polio, scabies, influenza, tuberculosis, and other human diseases are considered threats to nonhuman primates. The label for those infections is anthropozoonotic. Any of them might be carried by a tourist, a researcher, or a local person and have a potentially devastating impact on a tiny, isolated population of great apes with a relatively small gene pool, such as the mountain gorillas of Rwanda or the chimps of Gombe.

That's why Billy Karesh and his colleagues at the Wildlife Conservation Society label their program with the slogan "One World, One Health." The guiding principles come from ecology, of which human and veterinary medicine are merely subdisciplines. "It's not about wildlife health or about human health or about livestock health," Karesh told me. "There's really just one health" — the health and balance of ecosystems throughout the planet.

After our fruitless stakeout along the Mambili River in northwestern Congo, Karesh and I and Prosper Balo, along with other members of the team, traveled three hours downstream by pirogue. From there we drove along a dirt road to a town called Mbomo, the center point of an area where Ebola had killed 128 people during the same outbreak that struck gorillas at Lossi. We stopped at a little hospital, beside which stood a sign, painted in stark red letters:

ATTENTION EBOLA
NE TOUCHONS JAMAIS
NE MANIPULONS JAMAIS
LES ANIMAUX TROUVES
MORTS EN FORET
(Don't ever touch dead animals found in the forest.)

Mbomo was Balo's hometown. Visiting his house, we met his wife, Estelle, and some of his many children. We learned that Estelle's sister, two brothers, and another close relative all died of Ebola in 2003 and that Estelle herself was shunned by townspeople because of her association with the disease. No one would sell food to her. No one would touch her money. She had to hide in the forest. She would have died herself, Balo said, if he hadn't taught her the precautions he'd learned from Eric Leroy and the other scientists for whom he'd worked during the outbreak: sterilize everything with bleach, wash your hands, and don't touch corpses. But now the bad time was past and, with Balo's arm around her, Estelle was a smiling, healthy young woman.

Balo remembered the outbreak in his own way, mourning Estelle's losses and some of a different sort. He showed us a book, a botanical field guide, on the endpapers of which he had written a list of names: Apollo, Cassandra, Afrodita, and almost twenty others. They were gorillas, an entire group that he had known well, that he had tracked daily and observed lovingly at Lossi. Cassandra was his favorite, Balo said. Apollo was the silverback. "*Sont tous disparus en deux-mille trois,*" he said. All of them, gone in the 2003 outbreak. He'd lost his gorilla family and also members of his own family. It was very hard, Balo said.

For a long time he stood holding the book, opened for us to see those names. He comprehended emotionally what the scientists know from their data: That we — people and gorillas, horses and pigs and bats, monkeys and rats and mosquitoes and viruses — are all in this together.

RON ROSENBAUM

How to Trick an Online Scammer into Carving a Computer out of Wood

FROM *The Atlantic*

A VICIOUS AND INTRIGUING CYBERWAR has broken out in the spamosphere or, more specifically, in what I'd call the scamo-sphere.

I'm speaking of the emergence of "scam-baiters," the avengers of the scamosphere, who've arisen to take on "419" con artists, the scammers who pose in spam e-mails as agents for the widows of de-posed finance ministers of Dubai or vice chairmen of the Ivory Coast Cocoa Trading Board. The ones who promise you a share of a multimillion-dollar "inheritance" stashed in a Swiss bank account in return for your help in getting access to it by posing as the legal beneficiary. The ones who then try to persuade you (and it's amaz-ing how many are blinded enough by greed to believe the pitch) to fork over one "advance fee" after another to "estate attorneys," "private bank managers," and other fictional "facilitators" — until you awaken to the fact that you've been taken or are broke. (The name 419 comes from the section number of the Nigerian criminal code that applies to fraud, though the advance-fee fraud is actually a variation on the centuries-old "Spanish Prisoner" ploy.)

Scam-baiters have set out to reverse this dynamic, to turn the ta-bles on the scammers. The legions of scam-baiters seek to con the con artists, often with remarkable artistry of their own. They tease the scammers with promises of payments that don't arrive, with wired funds from banks that don't exist, with Western Union money transfers that go awry. They lead the scammers on wild-goose chases to pick up checks from couriers who don't materialize, insist that the scammers perform ridiculous stunts, and ask them to pose with demeaning signs to prove their commitment to the transaction. Blinded by the same greed that blinds their marks, the scammers take the scam-baiters' bait and, as often as not, end up as heads on the virtual wall in the scam-baiting websites' "trophy rooms."

The scam-baiters seem almost like a spontaneous evolutionary response to a threatening predatory species — think of them as the T cells of the Internet's immune system. But they can also seem an embodiment of the devolution of discourse and increase in abuse and invective that's come to be known as "cyber-disinhibition" — the tendency of people to engage in hostile interactions when they aren't inhibited by face-to-face contact.

Are the scam-baiters Jedi-like cyber-guardians taking up arms against the Web's Dark Side, spam-scam, or are they cyber-vigilantes engaging in vicious pranks that can at times border on racism?

On first entering the scam-baiting websites, one picks up the good-natured vibe of the elaborate fake bookie joints in movies such as *The Sting* — the hum and buzz of counter–con artists taking pleasure in the game. The chatter ranges from the relatively innocent-sounding "nov 7. i got somebody for the first time," with a transcript of a scam-bait string, to the more triumphalist "650 mile safari and longest insult EVER!"

"Safaris" are the trips scam-baiters lure scammers into making to remote banks to collect their advance fees, which, of course, don't exist. Insults, the bitter imprecations hurled at the scam-baiter once the scammer realizes he's been scammed, are prized as tokens of the baiter's success in "owning" the scammer — driving him around the bend and provoking him to the spluttering rage of capital-letter curses: "YOU ARE GOING STUPID, ARE YOU OUT OF YOUR MIND? YOU FOOLISH WHITE MONKEY AND YELLOW PIG" was the response of one "barrister" when finally copping to his humiliation. But the most valuable "trophies" are photos of scammers holding ridiculously worded signs — such as KING OF RETARDS or I AM A SHEEP SHAGGER — whose significance they apparently don't recognize.

*

I started paying closer attention to the world of 419 scams after a phone call from a woman I know. She nervously reported that out of curiosity she'd played along with a Nigerian scam and had just gotten an e-mail from someone called "The Professor" telling her that "diplomats" were on their way to her apartment with "documents to sign." I advised her to e-mail the Professor and tell him she'd called the police and the FBI. The diplomats never showed up.

Still, I'd never known anyone who'd gotten in that deep — though I'd certainly seen reports of the surprising success scammers have had with otherwise intelligent pillars of society, including a former congressman who'd served on the House committee weighing Nixon's impeachment for dirty tricks. My friend told me that she hadn't lost any funds thus far because she'd given them the number of a bank account with no money in it — which she realized wasn't exactly sufficient precaution.

I was fascinated to learn that she'd actually had a phone conversation with the Professor, a shadowy figure who shows up in a number of scam e-mails and whom I imagined as a kind of cyber-Moriarty — or perhaps a cyber-Virgil leading the unwary down into the lower circles of cyberscam hell. As I began paying more attention to cyberscam letters — to the subtle shifts in the pitch of the messages, to the tonal and rhetorical tropes — I began thinking of the vast body of these letters as a kind of literary genre.

I was particularly taken with the characters the scammers created: the widows and orphans of murdered dictators, the troubled bank managers, the associates of Russian oil billionaires hiding their wealth from Vladimir Putin. Here was food for literary exegesis: a sprawling international cast of recurring characters worthy, if not of *War and Peace,* then at least of Melville's *The Confidence-Man.* Like favorite characters in literature, they serve as our imaginary friends — which perhaps explains why so many lonely souls get conned by the phony plights of the scammers.

I realized that this literature and its subdomain of folk tales had evolving themes and memes. I noticed, for instance, a sudden epidemic of conscience-stricken "esophageal cancer" victims among the scam-letter writers, who in their "dying days" (after their disease had "defiled" all medical treatment, as one semiliterate appeal put it) had undergone conversions that had led them to ask for assistance in distributing huge but inaccessible fortunes to charity. Scam-lit had shifted its appeal from greed to altruism.

Then I saw a sudden proliferation of letters from "American soldiers" who'd found enormous hidden caches of Saddam Hussein's (or Uday's and Qusay's) ill-gotten gains and hoped (with your help, which of course would be amply rewarded) to shift the funds to an offshore account rather than turn them in to the authorities. Even those elusive weapons of mass destruction have shown up in this subgenre: in one e-mail a certain "Smith Scott," posing (I hope) as a Marine captain in Baghdad, claimed he'd discovered nuclear weapons in "some boxes"; he'd learned about them "in the process of tortur[ing]" some terrorists, who'd then led his troops to "a cave in Karbala."

One of the scam-baiters followed up with Captain Scott and, ignoring the boxes of money in the cave, expressed interest in the "nuclear devices." He received a disturbingly detailed reply:

These are complete nuclear weapons, RANGING FROM Mk-I TO Mk-III, NUCLEAR TYPE-BOMBS, WIDTH 28 AND 60.25 INCHES, LENGTH 42 AND 68 INCHES, WEIGHT 2,800 AND 3,400IB, Yields 15–16 Kt AND 18, 20–23, 37, 49 Kt RESPECTIVELY.

This specificity probably shouldn't be seen as evidence of possession of actual nuclear devices, though it might suggest where the "intelligence" on Saddam's WMD originated. What it does show is that the scamosphere follows the headlines.

I soon became riveted by the interaction between the scammers and the scam-baiting "community," particularly after discovering a frenetic hub of scam-baiters from all over the world, 419eater.com, with its explanation of the techniques of scam-baiting, its "mentor programme" for novices, and its intriguing philosophical discussions of the ethics of the counter-con.

The 419eater site was founded in October 2003 by Mike Berry, an Englishman who makes his living as an IT technician. It wasn't the first scam-baiting site, but it's now one of the largest and most active. "Back then, 2002, 2003, I was getting two or three of these scam letters in my inbox per week — now it's up to seventy-five or a hundred — and I would engage them in prolonged 'straight baits,'" Berry told me.

A "straight bait" is the entrapment of a scammer into an interminable correspondence that leads him to believe he's oh-so-close to getting his first advance fee for an inheritance transfer but is frustrated by one mistake, obstacle, and miscue after another until he finally realizes that *he's* the one being suckered. Unlike other scam-baits, which aim to embarrass the scammers, straight baits are meant to take up large amounts of scammers' time, keeping them from doing mischief. But the process can be tedious. Soon Berry and his cohorts at 419eater began attracting the more ambitious and imaginative among the scam-baiters of the world. (The site has now grown to some 20,000 registered members, and Berry estimates that about 10 percent of them are actively engaged in conning the con artists.) And there grew up a competition among the "elite baiters" and the "master baiters" (as they enjoy calling themselves) to see who could come up with the most elaborate and ridiculous ruses to engage the would-be thieves — so-called creative baits.

Over the past few years, Berry says, he's induced scammers to write out entire novels by hand, including the Lord of the Rings trilogy and most of the Harry Potter series; persuaded them to get tattoos (including one of the logo of the Holy Church of the Tattooed Saint); duped them into booking international flights and expensive hotel rooms to meet with his no-show personae; had them listen over the phone as Berry — who was supposedly just about to deliver that much-promised, many-times-delayed money transfer — faked his own death; and, mercilessly, made one of the scammers fall in love with Berry in his online guise as the actress Gillian Anderson.

I asked Berry how he managed to persuade a scammer to write out a Harry Potter novel.

"Well," Berry said, "he first wrote me with the usual Swiss-bank-account scam letter, and I then told him that I worked for a firm that did handwriting analysis, and we were looking for people to write out samples of their handwriting, and that we paid thirty dollars per page. Needless to say, he was eager to maximize his profit, and I suggested Harry Potter might be a good source. It must have kept him busy for a while."

And he must have been a bit disappointed when he shipped his manuscript off, never again to hear from the "handwriting analyst."

As time passed, Berry's creative engagements with the scammers became more and more elaborate. To my mind, his greatest achievement is the "Commodore 64" scam-bait, which I would not hesitate to call a scam-baiting work of art. In the hands of a master, a particularly ingenious, devious, and multilayered scam-bait is nothing less than an epistolary coup de théâtre.

The Commodore 64 counter-scam — which opens Berry's book *Greetings in Jesus Name!,* a compilation of baits and tips for newbies — began when Berry replied to an advance-fee-scam letter from an African who was "sitting on millions" and needed help getting access to it. Using the pseudonym "Derek Trotter" (after a British TV personality), Berry brushed aside the initial scam "deal" and claimed to represent an art gallery and foundation that sponsored promising sculptors. He suggested that if the scammer knew anyone interested in art, he should encourage him to apply for a scholarship.

When one "John Boko" (the same scammer, Berry believes) responded with interest a few days later, Trotter sent him elaborate specifications, along with a photograph of a cartoon cat and dog (from a British TV show) perched on a couch as the subject that had to be carved in wood — preferably wood that was "polished smooth" (the sign, of course, of all good art) — and sent to the UK to win the scholarship. Boko produced a weirdly convincing replica of the cat and dog. But when the expensive-to-ship carving arrived, Trotter claimed to be dismayed and disappointed to learn that it did not meet the precisely prescribed proportions, and in an e-mail to Boko, he suggested the sculpture might have suffered "shrinkage" during shipping. He accompanied his complaint, in a subsequent e-mail, with a trick photograph of the sculpture sitting next to a ruler to demonstrate that it didn't meet the specifications.

Trotter told Boko he regretted this unfortunate, unaccountable turn of events, but the foundation had strict specifications. Still, he thought that he could wangle him another chance. He offered Boko a special commission that he knew would earn him the scholarship his talent deserved. It turned out that Boko was still interested. He accepted a new challenge to produce a wood sculpture ("polished smooth") of a Commodore 64 computer, complete with raised keyboard and faithfully sculpted letters on the keys. The well-baited, totally hooked "sculptor" got back to work and succeeded in crafting a wooden replica of the Commodore. He even sent Trotter a photograph of it, a photo that has a Warholesque aura and makes the object seem like an exemplar of some yet-to-be-named genre — folk techno, maybe?

Once again Boko shipped off his highly polished work of art, and after a few delays, the package arrived. But after opening it, Trotter's "brother" Rodney, the "head of sales" at the gallery, sadly informed Boko, "It appears that your package was infiltrated by a rodent, more accurately a hamster" — the carefully sculpted Commodore of the photograph was now pocked with holes. (It doesn't quite look as if it's been gnawed by a hamster in the photo Trotter sent to Boko. One might suspect it had been bored with a drill.) Rodney wrote that the gnawed block of wood probably wouldn't qualify Boko for a scholarship, but said that he'd try to get an exception made.

The next e-mail Boko got was from the "UK Police," who reported that Derek Trotter had been arrested for fraud.

Brilliant. The Commodore 64 scam has a postmodern, meta feel to it, the counter-con being a piece of performance art *about* the creation of a work of art. Perhaps Boko didn't feel that way, but the beauty of this counter-scam artistry is that it wasn't done for money but for scripting (or con-scripting) a drama and for creating an object that's destined to last only a short time — all of which satisfy the criteria for a work of aesthetic contemplation, if not art.

Scam over, Berry posted pictures of the sculptures in the 419eater trophy room.

Ah yes, the trophy room. Trophies are what scam-baiters use to prove their mastery of the sport of exposing (and humiliating) scammers. The majority of the trophies at 419eater are photographs that the scammers (or their surrogates) have agreed to have taken of themselves, usually holding up signs, at the behest of the scam-baiters, who've convinced the scammers that for one reason or another the photos are essential to proceeding with the next step of the transaction, usually the transfer of some advance fee.

At times the trophy hunting on these scam-baiting sites can seem innocent in a clever con-artist way. But inside the trophy rooms, I found dozens of photos of black men and women wearing expressions that ranged from compliant to glum to humiliated to defiant

as they held up signs saying things such as I HAVE EXCESS VAGI-
NAL DISCHARGE. After looking at photo after photo, I felt uncomfortable — I'd lost any sense of vicarious victory over petty
thieves. It was like watching self-proclaimed Great White Hunters
abuse their beaters.

Some of the rhetoric of the scam-baiters is also troubling. They
boast of "owning" scammers, which carries unfortunate conno-
tations when it involves whites "owning" blacks. And there's the
mocking invective of the scam-baiting message boards and of the
trophy room, a place that I sense pricks the conscience of at least
some of the scam-baiters.

Or so it seems if one reads the "Ethics of Scambaiting" section on
the 419eater site. The site's webmasters take pains to disclaim racism, stating, "Racism is not permitted here . . . Individual posts are
reported for racism from time to time, and they are always acted on
speedily." The "ethics" statement argues that scam-baiters don't
pick their targets by race but that they also aren't constrained by
the race of their targets: they go after people of any race who seek
to cause loss, pain, and suffering to innocent victims. It claims that
there are a quarter of a million active scammers; that they cause
$1.5 billion in losses each year, an average of $20,000 per victim;
that the victims are not just people who want an illegal windfall but
people who think they're contributing to charities and orphan-
ages; that the victims have suffered more than financial losses, having been beaten, tortured, even murdered; and that the "degrading pictures" are sent voluntarily by scammers eager for ill-gotten
gains.

But then there are the counterarguments of the scammers, which
the ethics section both earnestly summarizes and attempts to refute. One scammer is quoted saying:

> I don't realy call it cheating . . . Some body has to pay what we call retri-
> bution From what Africa went through during the Slave trade era . . .
> The west took all our resourses, Manpower, and our cultural and tradi-
> tional wares . . . Some body will pay some how what your lineage owed

So scamming is retribution, reparation for imperialist exploitation. The fact that those colonialists who actually committed most
of the crimes are not the ones paying the reparations, that the
most helpless and naive members of the exploiter society (and, of

course, nonexploiter societies, since the scam letters go out across the globe and are as likely to snare a Yemenite as a Brooklynite) are paying the penalty doesn't come up. The scammers might counter that while their victims include many innocent people, the victims of colonialism included many *more* innocent people — many millions slaughtered, not just scammed.

One could make a sincere argument that the 419 scams exact a certain crude, if cruel and indiscriminate, justice for miseries caused by the West — if not justice, then justifiable vengeance. But of course greed, not justice or revenge, is the chief motive of the scammers. It's a complex issue.

The site's final word on racism: "Eater's anti-racism policy is sincere. We are genuinely offended by the accusation that we are racist, hence this effort to persuade you that we are not." A lot of explanation. Do the 419eaters protest too much?

I asked Mike Berry about the accusations of racism, and he said, "We're very careful about that. If you look at the pages of the trophy room, you see white faces too."

You certainly won't see many, and you will see lots of pages with only black faces. And sometimes the comments on the forum ridiculing the *mugus* ("fools," in Nigerian slang) seem to cross, or at least approach, the line between ridicule and abusive contempt, or at least condescension. While much of this can be seen as the good guys driving the bad guys crazy, cumulatively it has a disturbing quality.

The more I investigated the scam-baiting world, the more complex the questions and contradictions became. I learned, for instance, of a battalion of what you might call "interventionist" scam-baiters. They claim to use Internet service providers to identify a scammer's computer, then hack into it. They say that if they find a scam in progress, an innocent being led down the path to financial self-destruction by a con artist, they notify the victim that he's being played, disrupting the scammer's operation.

In other words, while scam-baiting usually involves some gray-area activities — impersonating people, manufacturing fake financial instruments — the interventionist scam-baiters are engaged in a darker gray area. And then there are troubling hints on the websites of an even darker practice known as "extreme anti-scamming," which seems to involve physical attacks on scammers. Are we losing track here of who's the criminal and who's the victim?

There also remains the question of cyber-disinhibition. Something about the lonely void in which online interactions are conducted seems to encourage the tendency toward the extreme or abusive mode in communication — because you're not face to face with the person you are berating or baiting. And the more I examined the scam-baiting community, the more troubled I was by evidence of this "online disinhibition effect," as psychologists call it.

I would like to give the scam-baiters the benefit of the doubt. At its best, scam-baiting can be seen as a kind of communal self-defense and a recompense for damage done — and it rarely involves physical harm or incarceration. (Scam-baiting may exact a toll in time and money and humiliation, but scammers largely escape official justice.) But when I look at the photos in the 419eater trophy room, I feel something has gone a bit wrong with the evolution of the scam-baiting community. What started out as a good-natured form of rough justice has become, in some respects, a theater of cruelty.

The more one investigates scam-baiting, the more one gets entangled in an emblematic ethical and behavioral question posed by the growth of communication in cyberspace. Are we getting a new, somewhat bleaker vision of human nature as we're freed from the bounds of real-time, face-to-face contact? Is the viciousness of the discourse what human nature would look like in a vacuum? Things certainly seem rancorous, not to mention regressive and infantile, in the gang wars of the right and left lobes of the political blogosphere, which (especially if you read the "comments") often seem more about humiliating and degrading those who take an opposing position than about persuading anyone of the rationality of one's arguments.

Does interaction at a distance disinhibit the urge to degrade? Is there some dynamic at work wherein the scamosphere calls forth the scam-baiters to do good deeds in response to bad deeds by scammers, but then the very doing of the deeds somehow draws forth a kind of deplorably vicious joy in inflicting psychic injury?

The competitive, exhibitionist escalation — the drive to come up with the most "creative" scam-bait — has become a drive to humiliate. Does this entail an irony — that as our civilization develops greater and greater technical sophistication in its ability to communicate, we find ourselves devolving into a cruder, Hobbesian state? When we contemplate the war between scammers and scam-baiters, must we say, "A pox on both their Webs"?

There's no denying the good that scam-baiters sometimes do. Superhero-like, they swoop in and save many innocents from ruin. But some in the scam-baiting community also take pleasure in mean-spirited mockery, like a mob of virtual vigilantes.

I have a modest suggestion that might make their methods (and their websites) more palatable: lose the trophy rooms. They're unnecessary, and they give what I truly hope is just a wrong impression — that scam-baiting is about the scam-baiter's ego, about triumphalism over the poor and uneducated.

I like my superheroes humble.

OLIVER SACKS

A Bolt from the Blue

FROM *The New Yorker*

TONY CICORIA WAS FORTY-TWO, very fit and robust, a former college football player who had become a well-regarded orthopedic surgeon in a small city in upstate New York. One afternoon in 1994, he was at a lakeside pavilion for a family gathering. It was pleasant and breezy, but he noticed a few storm clouds in the distance; it looked like rain.

He went to a pay phone outside the pavilion to make a quick call to his mother (this was before the age of cell phones). He still remembers every second of what happened next: "I was talking to my mother. There was a little bit of rain, thunder in the distance. My mother hung up. The phone was a foot away from where I was standing when I got struck. I remember a flash of light coming out of the phone. It hit me in the face. Next thing I remember, I was flying backwards."

Then — he seemed to hesitate before telling me this — "I was flying forwards. Bewildered. I looked around. I saw my own body on the ground. I said to myself, 'Oh shit, I'm dead.' I saw people converging on the body. I saw a woman — she had been waiting to use the phone right behind me — position herself over my body, give it CPR . . . I floated up the stairs — my consciousness came with me. I saw my kids, had the realization that they would be OK. Then I was surrounded by a bluish white light . . . an enormous feeling of well-being and peace. The highest and lowest points of my life raced by me. I had the perception of accelerating, being drawn up . . . There was speed and direction. Then, as I was saying to myself, 'This is the most glorious feeling I have ever had' — *slam!* I was back."

Cicoria knew that he was back in his own body because he felt pain — pain from the burns on his face and his left foot, where the electrical charge had entered and exited — and, he realized, "only bodies have pain." He wanted to go back, he wanted to tell the woman to stop giving him CPR, to let him go. But it was too late — he was firmly back among the living. After a minute or two, when he could speak, he said, "It's OK — I'm a doctor!" The woman, who turned out to be an intensive-care-unit nurse, replied, "A few minutes ago you weren't."

The police came and wanted to call an ambulance, but Cicoria refused. His family took him home ("It seemed to take hours," he recalled), where he called his own doctor, a cardiologist. The cardiologist thought that Cicoria must have had a brief cardiac arrest, but said, "With these things, you're alive or dead." He did not believe that Cicoria would suffer any aftereffects.

Cicoria also consulted a neurologist; he was feeling sluggish (most unusual for him) and having some difficulties with his memory. He found himself forgetting the names of people he knew well. He had a thorough neurological exam, an EEG, and an MRI. Again, nothing seemed amiss.

A couple of weeks later, when his energy returned, Cicoria went back to work. There were still some lingering memory problems — he occasionally forgot the names of rare diseases or surgical procedures — but his surgical skills were unimpaired. In another two weeks, his memory problems disappeared, and that, he thought, was the end of the matter.

What happened then fills Cicoria with amazement even now, a dozen years later. Life had returned to normal, seemingly, when "suddenly, over two or three days, there was this insatiable desire to listen to piano music." This was completely out of keeping with anything in his past. He had had a few piano lessons as a boy, he said, "but no real interest." He did not have a piano in his house. What music he did listen to tended to be rock-and-roll.

With this sudden onset of craving for piano music, he began to buy recordings and became especially enamored of a Vladimir Ashkenazy recording of Chopin favorites — the "Military" Polonaise, the "Winter Wind" Étude, the "Black Keys" Étude, the A-flat Major Polonaise. "I loved them all," Cicoria said. "I had the desire to play them. I ordered all the sheet music. At this point, one of our babysitters asked if she could store her piano in our house — so now, just when I craved one, a piano arrived, a nice little upright. It suited me fine. I could hardly read the music, could barely play, but I started to teach myself." It had been more than thirty years since the few piano lessons of his boyhood, and his fingers felt stiff and awkward.

And then, on the heels of this sudden desire for piano music, Cicoria started to hear music in his head. "The first time, it was in a dream," he said. "I was in a tux onstage; I was playing something I had written. I woke up, startled, and the music was still in my head. I jumped out of bed, started trying to write down as much of it as I could remember. But I hardly knew how to notate what I heard." This was not surprising — he had never tried to write or notate music before. But whenever he sat down at the piano to work on the Chopin, his own music "would come and take me over. It had a very powerful presence."

I was not quite sure what to make of this peremptory music, which would intrude and overwhelm him. Was he having musical hallucinations? No, Cicoria said, they were not hallucinations — "inspiration" was a more apt word. The music was there, deep inside him — or somewhere — and all he had to do was let it come to him. "It's like a frequency, a radio band. If I open myself up, it comes. I want to say, 'It comes from heaven,' as Mozart said." His music is ceaseless. "It never runs dry," he said.

Now he had to wrestle not just with learning to play Chopin but with giving form to the music in his head, trying it out on the piano, getting it down on manuscript paper. "It was a terrible struggle," he said. "I would get up at four in the morning and play till I went to work, and when I got home from work I was at the piano all evening. My wife was not really pleased. I was possessed."

In the third month after being struck by lightning, then, Cicoria — once an easygoing, genial family man, almost indifferent to music — was inspired, even possessed, by music, and scarcely had time for anything else. It began to dawn on him that perhaps he had been "saved" for a special purpose. "I came to think that the only reason I had been allowed to survive was the music," he said. I asked him whether he had been a religious man before the lightning. He had been raised Catholic, he said, but had never been particularly observant; he had some "unorthodox" ideas, too, such as a belief in reincarnation.

He himself, he grew to think, had had a sort of reincarnation —

had been transformed and given a special gift, a mission, to "tune in" to the music that he called, half metaphorically, "the music from heaven." This came, often, in "an absolute torrent" of notes with no breaks, no rests between them, and he would have to give it shape and form. (As he said this, I thought of Caedmon, the seventh-century Anglo-Saxon poet, an uneducated herdsman who, it was said, had received the art of song in a dream one night and spent the rest of his life praising God and creation in hymns and poems.)

Cicoria continued to work on his piano technique and his compositions. He would travel to concerts by his favorite performers but had nothing to do with musical friends or musical activities in his own town. This was a solitary pursuit, between himself and his muse.

I asked whether he had experienced other changes since the lightning strike — a new appreciation of art, perhaps, different taste in reading, new beliefs. Cicoria said he had become "very spiritual." He had started to read every book that he could find about near-death experiences and about lightning strikes. And he had acquired "a whole library on Tesla," as well as anything on the terrible and beautiful power of high-voltage electricity. He thought he could sometimes feel "auras" of light or energy around people's bodies — he had never felt this before the lightning bolt.

Some years passed, and Cicoria's new life, his inspiration, never deserted him. He continued to work full time as a surgeon, but his heart and mind now centered on music. He got divorced in 2004, and that same year he had a fearful motorcycle accident. He had no memory of it, but his Harley was struck by another vehicle and he was found unconscious in a ditch, with broken bones, a ruptured spleen, a perforated lung, cardiac contusions, and, despite his helmet, head injuries. Nonetheless, he made a complete recovery and was back at work in two months. Neither the accident nor his head injury nor his divorce seemed to have made any difference to his passion for playing and composing music.

I have never met another person with a story like Tony Cicoria's, but I have occasionally had patients with a similar sudden onset of musical or artistic interests — including a research chemist in her early forties, Salimah M. (I have changed her name and some identifying details.) In 2002 Salimah started to have brief episodes, lasting a minute or less, in which she would get "a strange feeling" — sometimes a sense that she was on a beach that she had once known while at the same time being perfectly conscious of her current surroundings and able to continue a conversation, or drive a car, or do whatever she had been doing. Occasionally, the episodes were accompanied by a "sour taste" in her mouth. She noticed these strange occurrences but did not think of them as having any neurological significance. It was only when she had a grand mal seizure in the summer of 2003 that she went to a neurologist and was given brain scans, which revealed a large tumor in her right temporal lobe — the cause of her peculiar episodes. The tumor, her doctors felt, was malignant (though it was probably an oligodendroglioma, of relatively low malignancy) and needed to be removed. Salimah wondered if she had been given a death sentence and was fearful of the operation and its possible consequences; she and her husband had been told that it might cause some "personality changes." But in the event the surgery went well, most of the tumor was removed, and, after a period of convalescence, Salimah was able to return to her work as a chemist.

Before the surgery, Salimah had been a fairly reserved woman who would occasionally be annoyed or preoccupied by small things like dust or untidiness; her husband said that she was sometimes "obsessive" about jobs that needed to be done around the house. But now, after the surgery, she seemed unperturbed by such domestic matters. She had become, in the idiosyncratic words of her husband (English was not their first language), "a happy cat." She was, he declared, "a joyologist."

Salimah's new cheerfulness was apparent at work. She had worked in the same laboratory for fifteen years and had always been admired for her intelligence and dedication. Yet while losing none of this professional competence, she seemed a much warmer person, keenly sympathetic and interested in the lives and feelings of her coworkers. Where before, in a colleague's words, she had been "much more into herself," she now became the confidante and social center of the entire lab.

At home, too, she shed some of her Marie Curie–like, work-oriented personality. She permitted herself time off from her thinking, her equations, and became more interested in going to movies and parties, living it up a bit. And a new love, a new passion, entered her life. As a girl she had been, in her own words, "vaguely musical," had played the piano a little, but music had never played any great part in her life. Now it was different. She longed to go to concerts, to listen to classical music on the radio or on CDs. She could be moved to rapture or tears by music that had carried "no special feeling" for her before. She became "addicted" to her car radio, which she would listen to while driving to work. A colleague who happened to pass Salimah in her convertible on the road said that the music on her radio was "incredibly loud" — he could hear it a quarter of a mile away. Salimah was "entertaining the whole freeway."

Like Tony Cicoria, Salimah showed a drastic transformation from being only vaguely interested in music to being passionately excited by it and in continual need of it. And with both of them there were other, more general changes, too — a surge of emotionality, as if emotions of every sort were being stimulated or released. In Salimah's words, "What happened after the surgery — I felt reborn. That changed my outlook on life and made me appreciate every minute of it."

Could someone develop a "pure" musicophilia, without any accompanying changes in personality or behavior? In 2006 just such a situation was described by the neurologists Jonathan Rohrer, Shelagh Smith, and Jason Warren in their striking case history of a Thai woman in her midsixties who had severe temporal-lobe epilepsy. After seven years, her seizures were finally brought under control by the anticonvulsant drug lamotrigine (LTG). Prior to starting on this medication, as Rohrer and his colleagues wrote,

> she had always been indifferent to music, never listening for pleasure or attending concerts. This was in contrast to her husband and daughter, who played the piano and the violin . . . She was unmoved by the traditional Thai music she had heard at family and public events in Bangkok and by classical and popular genres of Western music after she moved to the United Kingdom. Indeed, she continued to avoid music where possible, and actively disliked certain musical timbres (for example, she would shut the door to avoid hearing her husband playing piano music, and found choral singing "irritating").

This changed abruptly when the patient was put on lamotrigine, they continued:

> Within several weeks of starting LTG, [she] sought out musical programmes on the radio and television, listened to classical-music stations on the radio for many hours each day, and demanded to attend concerts. Her husband described how she had sat "transfixed" throughout "La Traviata" and became annoyed when other audience members talked during the performance. She now described listening to classical music as an extremely pleasant and emotion-charged experience. She did not sing or whistle, and no other changes were found in her behavior or personality. No evidence of thought disorder, hallucinations, or disturbed mood was seen.

While Rohrer and his colleagues could not pinpoint the precise basis of their patient's musicophilia, they hazarded the suggestion that during her years of seizure activity, she might have developed an intensified neural connection between perceptual systems in the temporal lobes and certain parts of the limbic system involved in emotional response — a connection that became apparent only when her seizures were brought under control with medication. In the 1970s the neuropsychiatrist David M. Bear suggested that such a sensory-limbic hyperconnection might be the basis for the emergence of the unexpected artistic, sexual, mystical, or religious feelings that sometimes occur in people with temporal-lobe epilepsy. Could something similar have occurred with Tony Cicoria?

Last spring Cicoria took part in a ten-day retreat for music students, gifted amateurs, and professionals. The music camp doubles as a showroom for Erica vanderLinde Feidner, a concert pianist who also specializes in finding the perfect piano for each of her clients. Cicoria bought one of her pianos there — a Bösendorfer grand, a unique prototype made in Vienna. She thought he had a remarkable instinct for picking out a piano with exactly the tone he wanted. It was, Cicoria felt, a good time — a good place — to make his debut as a musician.

He prepared two pieces for his concert: his first love, Chopin's B-flat Minor Scherzo, and his own first composition, which he called Rhapsody, Opus 1. His playing, and his story, electrified everyone at the retreat (many expressed the fantasy that they, too, might be struck by lightning). He played, Erica said, with "great passion,

great brio," and, if not with genius, at least with creditable skill —
an astounding feat for someone with virtually no musical background who, at forty-two, had taught himself to play.

What, in the end, did I think of his story? Dr. Cicoria asked me.
Had I ever encountered anything similar? I asked him what *he*
thought, and how he would interpret what had happened to him.
He replied that as a medical man he was at a loss to explain these
events, and he had to think of them in "spiritual" terms. I coun-
tered that with no disrespect to the spiritual, I felt that even the
most exalted states of mind — the most extraordinary transformations — must have some physical basis or at least some physiologi-
cal correlate in neural activity.

At the time of his lightning strike, Cicoria had both a near-death
experience and an out-of-body experience. Many supernatural or
mystical explanations have arisen to explain out-of-body experiences, but they have also been a topic of neurological investigation for a century or more. Such experiences are relatively stereo-
typed in format: one seems to be no longer in one's own body but
outside it and, most commonly, looking down on oneself from
eight or nine feet above (neurologists refer to this as autoscopy).
One seems to see clearly the room and people and objects nearby,
but from an aerial perspective. People who have had such experiences often describe vestibular sensations like "floating" or "flying." Out-of-body experiences can inspire fear or joy or a feeling of
detachment, but they are usually described as intensely "real" —
not at all like a dream or a hallucination. They have been reported
in many sorts of near-death experiences, as well as in temporal-lobe
seizures. There is some evidence that both the visuospatial and vestibular aspects of out-of-body experiences are related to disturbed
function in the cerebral cortex, especially at the junctional region
between the temporal and parietal lobes. (The neurologist Orrin
Devinsky and others have described "autoscopic phenomena with
seizures" in ten of their patients and have reviewed similar cases in
the medical literature, while Olaf Blanke and his colleagues in Swit-
zerland have been able to monitor the brain activity of epileptic patients actually undergoing out-of-body experiences.)

But it was not just an out-of-body experience that Dr. Cicoria reported. He saw a bluish white light, he saw his children, his life

flashed past him, he had a sense of ecstasy, and, above all, he had a sense of something transcendent and enormously significant. What could be the neural basis of this? Similar near-death experiences have often been described by people who have been or believed themselves to be in great danger, whether they were involved in sudden accidents, or struck by lightning, or, most commonly, revived after cardiac arrest. All of these are situations not only fraught with terror but likely to cause a sudden drop in blood pressure and cerebral blood flow. There is likely to be intense emotional arousal and a surge of noradrenaline and other neurotransmitters in such states, whether the effect is one of terror or rapture. We have, as yet, little idea of the actual neural correlates of such experiences, but the alterations of consciousness and emotion that occur are very profound and must involve the emotional parts of the brain — including the amygdala and brain-stem nuclei — as well as the cortex. (Kevin R. Nelson and his colleagues at the University of Kentucky have published several papers stressing the similarities between the dissociation, euphoria, and mystical feelings of near-death experiences and those of dreaming, REM sleep, and the hallucinatory states in the borderlands of sleep.)

While out-of-body experiences have the character of a perceptual illusion (albeit a complex and singular one), near-death experiences have all the hallmarks of mystical experience, as William James defines it: passivity, ineffability, transience, and a noetic quality. In a near-death experience, one is totally consumed — swept up, almost literally, in a blaze (sometimes a tunnel or funnel) of light, and drawn toward a Beyond — beyond life, beyond space and time. There is a sense of a last look, a (greatly accelerated) farewell to things earthly, the places and people and events of one's life, and a sense of ecstasy or joy as one soars toward one's destination. Experiences like this are not easily dismissed by those who have been through them, and they may sometimes lead to a conversion or metanoia, a change of mind, that alters the direction and orientation of a life. One cannot suppose, any more than one can with out-of-body experiences, that such events are pure fancy; similar features are emphasized in every account. Near-death experiences must also have a neurological basis of their own, one that profoundly alters consciousness itself.

What about Tony Cicoria's remarkable access of musicality, his sudden musicophilia? Patients with degeneration of the front parts of the brain, so-called frontotemporal dementia, sometimes develop a startling emergence or release of musical talents and passions as they lose the powers of abstraction and language — but clearly this was not the case with Cicoria, who was articulate and highly competent in every way. In 1984 Daniel Jacome described a patient who had had a stroke that damaged the left hemisphere of his brain and who subsequently developed "hypermusia" and "musicophilia," along with aphasia and other problems. But there was nothing to suggest that Cicoria had experienced any significant brain damage after the lightning strike, other than a very transient disturbance to his memory systems for a week or two.

His situation did remind me a bit of that of Franco Magnani, the "memory artist" of whom I have written. Magnani had never thought of being a painter until he experienced a strange crisis or illness — perhaps a form of temporal-lobe epilepsy — when he was thirty-one. He had nightly dreams of Pontito, the little Tuscan village where he was born; after he woke, these images remained intensely vivid, with full depth and reality ("like holograms"). Magnani was consumed by a need to make these images real, to paint them, and so he taught himself to paint, devoting every free minute to producing hundreds of views of Pontito.

Could Tony Cicoria's musical dreams, his musical inspirations, have been epileptic in nature? Such a question cannot be answered with a simple EEG such as Cicoria had following his accident but would require special EEG monitoring over the course of many days.

And why was there such a delay in the development of Cicoria's musicophilia? What was happening in the six or seven weeks that elapsed between his cardiac arrest and the rather sudden eruption of musicality? We know that there were temporary aftereffects — the confused state that lasted for a few hours and the disturbance of memory that went on for a couple of weeks. These could have been due to cerebral anoxia alone — for his brain must have been without adequate oxygen for a minute or more. One has to suspect, however, that Cicoria's apparent recovery was not as complete as it seemed and that his brain was still reacting to the original insult and reorganizing itself during this time.

Cicoria feels that he is "a different person" now — musically, emotionally, psychologically, and spiritually. This was my impression, too, as I listened to his story and saw something of the new passions that had transformed him. Looking at him from a neurological vantage point, I suspected that his brain must be very different now from what it had been before he was hit by lightning or in the days immediately following, when neurological tests showed nothing grossly amiss. Could we now, a dozen years later, define the neurological basis of his musicophilia? Many new and far subtler tests of brain function have been developed since Cicoria had his injury, and he agreed that it would be interesting to investigate this further. But after a moment he reconsidered and said that perhaps it was best to let things be. His was a lucky strike, and the music, however it had come, was a blessing, a grace — not to be questioned.

MICHAEL SPECTER

Darwin's Surprise

FROM *The New Yorker*

THIERRY HEIDMANN'S OFFICE, adjacent to the laboratory he runs at the Institut Gustave Roussy, on the southern edge of Paris, could pass for a museum of genetic catastrophe. Files devoted to the world's most horrifying infectious diseases fill the cabinets and line the shelves. There are thick folders for smallpox, Ebola virus, and various forms of influenza. SARS is accounted for, as are more obscure pathogens, such as feline leukemia virus, Mason-Pfizer monkey virus, and simian foamy virus, which is endemic in African apes. HIV, the best known and most insidious of the viruses at work today, has its own shelf of files. The lab's beakers, vials, and refrigerators, secured behind locked doors with double-paned windows, all teem with viruses. Heidmann, a meaty, middle-aged man with wild eyebrows and a beard heavily flecked with gray, has devoted his career to learning what viruses might tell us about AIDS and various forms of cancer. "This knowledge will help us treat terrible diseases," he told me, nodding briefly toward his lab. "Viruses can provide answers to questions we have never even asked."

Viruses reproduce rapidly and often with violent results, yet they are so rudimentary that many scientists don't even consider them to be alive. A virus is nothing more than a few strands of genetic material wrapped in a package of protein — a parasite, unable to function on its own. In order to survive, it must find a cell to infect. Only then can any virus make use of its single talent, which is to take control of a host's cellular machinery and use it to churn out thousands of copies of itself. These viruses then move from one cell to the next, transforming each new host into a factory that makes even more virus. In this way, one infected cell soon becomes billions.

Nothing — not even plague — has posed a more persistent threat to humanity than viral diseases: yellow fever, measles, and smallpox have been causing epidemics for thousands of years. At the end of World War I, 50 million people died of Spanish flu; smallpox may have killed half a billion during the twentieth century alone. Those viruses were highly infectious, yet their impact was limited by their ferocity: a virus may destroy an entire culture, but if we die it dies, too. As a result, not even smallpox possessed the evolutionary power to influence humans as a species — to alter our genetic structure. That would require an organism to insinuate itself into the critical cells we need in order to reproduce: our germ cells. Only retroviruses, which reverse the usual flow of genetic code from DNA to RNA, are capable of doing that. A retrovirus stores its genetic information in a single-stranded molecule of RNA instead of the more common double-stranded DNA. When it infects a cell, the virus deploys a special enzyme, called reverse transcriptase, that enables it to copy itself and then paste its own genes into the new cell's DNA. It then becomes part of that cell forever; when the cell divides, the virus goes with it. Scientists have long suspected that if a retrovirus happens to infect a human sperm cell or egg, which is rare, and if that embryo survives — which is rarer still — the retrovirus could take its place in the blueprint of our species, passed from mother to child and from one generation to the next, much like a gene for eye color or asthma.

When the sequence of the human genome was fully mapped in 2003, researchers discovered something they had not anticipated: our bodies are littered with shards of such retroviruses, fragments of the chemical code from which all genetic material is made. It takes less than 2 percent of our genome to create all the proteins necessary for us to live; 8 percent, however, is composed of broken and disabled retroviruses, which, millions of years ago, managed to embed themselves in the DNA of our ancestors. They are called endogenous retroviruses, because once they infect the DNA of a species they become part of that species. One by one, though, after molecular battles that raged for thousands of generations, they have been defeated by evolution. Like dinosaur bones, these viral fragments are fossils. Instead of having been buried in sand, they reside within each of us, carrying a record that goes back millions of years. Because they no longer seem to serve a purpose or cause harm, these remnants have often been referred to as "junk DNA." Many still manage to generate proteins, but scientists have never found one that functions properly in humans or that could make us sick.

Then, last year, Thierry Heidmann brought one back to life. Combining the tools of genomics, virology, and evolutionary biology, he and his colleagues took a virus that had been extinct for hundreds of thousands of years, figured out how the broken parts were originally aligned, and then pieced them together. After resurrecting the virus, the team placed it in human cells and found that their creation did indeed insert itself into the DNA of those cells. They also mixed the virus with cells taken from hamsters and cats. It quickly infected them all, offering the first evidence that the broken parts could once again be made infectious. The experiment could provide vital clues about how viruses like HIV work. Inevitably, though, it also conjures images of Frankenstein's monster and *Jurassic Park*.

"If you think about this for five minutes, it is wild stuff," John Coffin told me when I visited him in his laboratory at Tufts University, where he is the American Cancer Society Research Professor. Coffin is one of the country's most distinguished molecular biologists and was one of the first to explore the role of endogenous retroviruses in human evolution. "I understand that the idea of bringing something dead back to life is fundamentally frightening," he went on. "It's a power that science has come to possess and it makes us queasy, and it should. But there are many viruses that are more dangerous than these — more infectious, far riskier to work with, and less potentially useful."

Thanks to steady advances in computing power and DNA technology, a talented undergraduate with a decent laptop and access to any university biology lab can assemble a virus with ease. Five years ago, as if to prove that point, researchers from the State University of New York at Stony Brook "built" a polio virus, using widely available information and DNA they bought through the mail. To test their "polio recipe," they injected the virus into mice. The animals first became paralyzed and then died. ("The reason we did it was to prove that it can be done," Eckard Wimmer, who led the team, said at the time. "Progress in biomedical research has its benefits and it has its downside.") The effort was widely seen as pointless and the justification absurd. "Proof of principle for bio-terrorism," Coffin called it. "Nothing more." Then two years ago, after researchers had sequenced the genetic code of the 1918 flu virus, federal scientists reconstructed it, too. In that case, there was a well-understood and highly desired goal: to develop a vaccine that might offer protection against future pandemics.

Resurrecting an extinct virus is another matter. Still, if Heidmann had stuck to scientific nomenclature when he published his results in the fall of 2006, few outside his profession would have noticed. A paper entitled "Identification of an Infectious Progenitor for the Multiple-Copy HERV-K Human Endogenous Retroelements," which appeared in the journal *Genome Research,* was unlikely to cause a stir. Heidmann is on a bit of a mission, though. He named the virus Phoenix, after the mythical bird that rises from the ashes, because he is convinced that this virus and others like it have much to tell about the origins and evolution of humanity.

With equal ardor but less fanfare, scientists throughout the world have embarked on similar or related projects. One team, at the Aaron Diamond AIDS Research Center in New York, recently created an almost identical virus. In the past few months, groups at Oxford University and at the Fred Hutchinson Cancer Research Center in Seattle have also produced results that provide startling observations about evolution and disease. The approaches often differ, but not the goals. All of these researchers hope that excavating the molecular past will help address the medical complexities that we confront today. Almost incidentally, they have created a new discipline, paleovirology, which seeks to better understand the impact of modern diseases by studying the genetic history of ancient viruses.

"This is something not to fear but to celebrate," Heidmann told me one day as we sat in his office at the institute, which is dedicated to the treatment and eradication of cancer. Through the window, the Eiffel Tower hovered silently over the distant city. "What is remarkable here, and unique, is the fact that endogenous retroviruses are two things at once: genes and viruses. And those viruses helped make us who we are today just as surely as other genes did. I am not certain that we would have survived as a species without them."

He continued, "The Phoenix virus sheds light on how HIV operates but, more than that, on how *we* operate and how we evolved. Many people study other aspects of human evolution — how we came to walk or the meaning of domesticated animals. But I would argue that equally important is the role of pathogens in shaping the way we are today. Look, for instance, at the process of pregnancy and birth." Heidmann and others have suggested that without endogenous retroviruses mammals might never have developed a placenta, which protects the fetus and gives it time to mature. That led to live birth, one of the hallmarks of our evolutionary success over birds, reptiles, and fish. Eggs cannot eliminate waste or draw the maternal nutrients required to develop the large brains that have made mammals so versatile. "These viruses made those changes possible," Heidmann told me. "It is quite possible that without them, human beings would still be laying eggs."

HIV, the only retrovirus that most people have heard of, has caused more than 25 million deaths and infected at least twice that number of people since the middle of the twentieth century, when it moved from monkey to man. It may be hard to understand how organisms from that same family, constructed with the same genes, could have played a beneficial, and possibly even essential, role in the health and development of any species. In 1968 Robin Weiss, who is now a professor of viral oncology at University College London, found endogenous retroviruses in the embryos of healthy chickens. When he suggested that they were not only benign but might actually perform a critical function in placental development, molecular biologists laughed. "When I first submitted my results on a novel 'endogenous' envelope, suggesting the existence of an integrated retrovirus in normal embryo cells, the manuscript was roundly rejected," Weiss wrote last year in the journal *Retrovirology*. "One reviewer pronounced that my interpretation was impossible." Weiss, who is responsible for much of the basic knowledge about how the AIDS virus interacts with the human immune system, was not deterred. He was eager to learn whether the chicken retroviruses he had seen were recently acquired infections or inheritances that had been passed down through the centuries.

He moved to the Pahang jungle of Malaysia and began living with a group of Orang Asli tribesmen. Red jungle fowl, an ancestor species of chickens, were plentiful there, and the tribe was skilled at trapping them. After collecting and testing both eggs and blood samples, Weiss was able to identify versions of the same viruses. Similar tests were soon carried out on other animals. The discovery helped mark the beginning of a new approach to biology. "If Charles Darwin reappeared today, he might be surprised to learn that humans are descended from viruses as well as from apes," Weiss wrote.

Darwin's surprise almost certainly would be mixed with delight: when he suggested, in *The Descent of Man* (1871), that humans and apes shared a common ancestor, it was a revolutionary idea, and it remains one today. Yet nothing provides more convincing evidence for the "theory" of evolution than the viruses contained within our DNA. Until recently, the earliest available information about the history and course of human diseases, like smallpox and typhus, came from mummies no more than 4,000 years old. Evolution cannot be measured in a time span that short. Endogenous retroviruses provide a trail of molecular bread crumbs leading millions of years into the past.

Darwin's theory makes sense, though, only if humans share most of those viral fragments with relatives like chimpanzees and monkeys. And we do, in thousands of places throughout our genome. If that were a coincidence, humans and chimpanzees would have had to endure an incalculable number of identical viral infections in the course of millions of years, and then somehow those infections would have had to end up in exactly the same place within each genome. The rungs of the ladder of human DNA consist of 3 billion pairs of nucleotides spread across forty-six chromosomes. The sequences of those nucleotides determine how each person differs from another and from all other living things. The only way that humans, in thousands of seemingly random locations, could possess the exact retroviral DNA found in another species is by inheriting it from a common ancestor.

Molecular biology has made precise knowledge about the nature of that inheritance possible. With extensive databases of genetic sequences, reconstructing ancestral genomes has become common, and retroviruses have been found in the genome of every vertebrate species that has been studied. Anthropologists and biologists have used them to investigate not only the lineage of primates but the relationships among animals — dogs, jackals, wolves, and foxes, for example — and also to test whether similar organisms may in fact be unrelated.

Although it is no longer a daunting technical task to find such viruses or their genes, figuring out the selective evolutionary pressures that shaped them remains difficult. Partly that is because the viruses mutate with such speed. HIV can evolve a million times as fast as the human immune-system cells it infects. (Such constant change makes it hard to develop antiviral drugs that will remain effective for long, and it has also presented a significant obstacle to the development of an AIDS vaccine.)

There are retroviruses (like HIV) that do not infect sperm or egg cells. Because they are not inherited, they leave no trace of their history. "We can have a fossil record only of the viruses that made it into the germ line," Paul Bieniasz told me. "And, of course, most did not." Bieniasz is a professor of retrovirology at the Aaron Diamond AIDS Research Center and the chief of the retrovirology laboratory at Rockefeller University. He has long been interested in the way complex organisms interact with viruses and adapt to them. "With flu virus, you can watch it change in real time," he said. "You can watch the antibodies develop and see when and how it dies out. But with these others you are looking back tens of millions of years, so it is hard to know how a virus functioned."

While Heidmann was working with the Phoenix virus in France, Bieniasz and two colleagues at Aaron Diamond initiated a similar project. (At first, neither team was aware of the other's work.) Bieniasz rebuilt the youngest extinct retrovirus in the human genome — one that was still active a few hundred thousand years ago — because it had the fewest mutations. The team took ten versions of that virus (we carry more than thirty) and compared the thousands of nucleotides in the genetic sequence of each version. They were almost identical, but where they differed the researchers selected the nucleotides that appeared most frequently. That permitted them to piece together a working replica of the extinct retrovirus. "If you have a person with a lethal defect in the heart," Bieniasz explained, "and another with a lethal defect in the kidney, you could make one healthy person by transplanting the respective organs. That is what we did.

"In the past, you got sick and you keeled over and died," he said. "Or you survived. Nobody could make much sense of it. But almost ten percent of our DNA consists of old retroviruses, and that says to me that it's pretty clear they played a major role in our evolution. We evolved remarkably sophisticated defenses against them, and we would have done that only if their impact on human populations had been quite severe. It's very likely that we have been under threat from retroviruses many times throughout human history. It is eminently possible that this is not the first time we have been colonized by a virus very much like HIV."

At the end of the nineteenth century, a mysterious series of cancer epidemics devastated American poultry farms. One bird would fall ill and the entire flock would soon be dead. In 1909 a desperate farmer from Long Island brought a chicken with a tumor to the laboratory of Peyton Rous, a young cancer researcher at the Rockefeller Institute for Medical Research (which became Rockefeller University) in New York City. Cancer was not supposed to spread by virus, but the bird clearly had cancer. Rous, who as a young man had worked on a Texas cattle ranch, was mystified. He extracted cancer cells from the sick bird, chopped them up, and injected the filtered remains into healthy chickens: they all developed tumors. A virus had to be the cause, but for years no one could figure out how the virus functioned.

Then in the 1960s, Howard Temin, a virologist at the University of Wisconsin, began to question the "central dogma" of molecular biology, which stated that genetic instructions moved in a single direction, from the basic blueprints contained within our DNA to RNA, which translates those blueprints and uses them to build proteins. He suggested that the process could essentially run in the other direction: an RNA tumor virus could give rise to a DNA copy, which would then insert itself into the genetic material of a cell. Temin's theory was dismissed, as most fundamental departures from conventional wisdom are. But he never wavered. Finally in 1970, he and David Baltimore, who was working in a separate lab, at the Massachusetts Institute of Technology, simultaneously discovered reverse transcriptase, the special enzyme that can do exactly what Temin predicted: make DNA from RNA.

The discovery has had a profound impact on modern medicine. It not only explained how cancer can be caused by a virus but provided researchers with the tools they needed to understand the origins and natural progression of diseases like AIDS. It also created a new field, retrovirology, and, more than that, as the Nobel committee noted when it awarded the 1975 prize in medicine to both Baltimore and Temin, it began to erase the tenuous borders between viruses and genes.

Retroviruses cause cancers in chickens, sheep, mice, and other animals, but their effect on humans became clear only in the late 1970s with the identification of two viruses that cause forms of leukemia. Retroviral proteins are particularly abundant in certain kinds of tumor cells, and scientists wondered to what degree they might be a cause of cancer. They were also curious about how retroviruses that infect us today differ from their ancestors. Working with mice in 2005, Thierry Heidmann found that endogenous retroviruses were present in large quantities in tumor cells. Similar viruses have been associated with many cancers (and other diseases). It is still not clear how they function, but they may help subvert the immune system, which would permit cancer cells to grow without restraint. Endogenous retroviruses also may actually protect us from viruses that are even worse. Experiments with mice and chickens have shown that they can block new infections by viruses with a similar genetic structure. Nonetheless, endogenous retroviruses are parasites, and in most cases the cells they infect would be better off without them. There is, however, one notable exception.

The earliest mammals, ancestors of the spiny anteater and the duck-billed platypus, laid eggs. Then, at least 100 million years ago, embryos, instead of growing in a shell, essentially became parasites. While still only balls of cells, they began to implant themselves in the lining of the womb. The result was the placenta, which permits the embryos to take nourishment from the mother's blood while preventing immune cells or bacteria from entering. The placenta is essentially a modified egg. In the early 1970s, biologists who were scanning baboon placentas with an electron microscope were surprised to see retroviruses on a layer of tissue known as the syncytium, which forms the principal barrier between mother and fetus. They were even more surprised to see that all the animals were healthy. The same phenomenon was soon observed in mice, cats, guinea pigs, and humans. For many years, however, embryologists were not quite sure what to make of these placental discoveries. Most remained focused on the potential harm a retrovirus could cause rather than on any possible benefit. Cell fusion is a fundamental characteristic of the mammalian placenta but also, it turns out, of endogenous retroviruses. In fact, the protein syncytin, which causes placental cells to fuse together, employs the exact mechanism that enables retroviruses to latch on to the cells they infect.

The Nobel Prize–winning biologist Joshua Lederberg once wrote that the "single biggest threat to man's continued dominance on this planet is the virus." Harmit Malik, an evolutionary geneticist at the Fred Hutchinson Cancer Research Center, acknowledges the threat, yet he is confident that viruses may also provide one of our greatest scientific opportunities. Exploring that fundamental paradox — that our most talented parasites may also make us stronger — has become Malik's passion. "We have been in an evolutionary arms race with viruses for at least one hundred million years," he told me recently when I visited his laboratory. "There is genetic conflict everywhere. You see it in processes that you would never suspect — in cell division, for instance, and in the production of proteins involved in the very essence of maintaining life.

"One party is winning and the other losing all the time," Malik went on. "That's evolution. It's the world's definitive game of cat and mouse. Viruses evolve, the host adapts, proteins change, viruses evade them. It never ends." The AIDS virus, for example, has one gene, called vif, that does nothing but block a protein whose sole job is to stop the virus from making copies of itself. It simply takes that protein into the cellular equivalent of a trash can; if not for that gene, HIV might have been a trivial disease. "To even think about the many million-year processes that caused that sort of evolution," Malik said, shaking his head in wonder. "It's dazzling." Malik grew up in Bombay and studied chemical engineering there at the Indian Institute of Technology, one of the most prestigious technical institutions in a country obsessed with producing engineers. He gave no real thought to biology, but he was wholly uninspired by his other studies. "It was fair to say I had little interest in chemical engineering, and I happened to tell that to my faculty adviser," he recalled. "He asked me what I liked. Well, I was reading

Richard Dawkins at the time, his book *The Selfish Gene*" — which asserts that a gene will operate in its own interest even if that means destroying an organism that it inhabits or helped create. The concept fascinated Malik. "I was thinking of becoming a philosopher," he said. "I thought I would study selfishness."

Malik's adviser had another idea. The university had just established a department of molecular biology, and Malik was dispatched to speak with its director. "This guy ended up teaching me by himself, sitting across the table. We met three times a week. I soon realized that he was testing out his course on me. I liked it and decided to apply to graduate school — although I had less than a tenth of the required biology courses. I had very little hope." But he had excellent test scores and in 1993 was accepted at the University of Rochester as a graduate student in the biology department. He visited his new adviser as soon as he arrived. "He looked at my schedule and said, 'I see that you are doing genetics.' I had no clue what he was talking about, but I said sure, that sounds good. I had never taken a course in the subject. He gave me the textbook and told me that the class was for undergraduates, which made me feel more comfortable." It wasn't until the end of the conversation that Malik realized he would be teaching the class, not taking it.

The Hutchinson Center encourages its research scientists to collaborate with colleagues in seemingly unrelated fields. Malik and Michael Emerman, a virologist at the center's Human Biology and Basic Sciences Divisions, have been working together for four years. Malik's principal interest is historical: why did evolutionary pressures shape our defenses against viruses, and how have they done it? Emerman studies the genetic composition and molecular pathology of the AIDS virus. "Together we are trying to understand what constellation of viruses we are susceptible to and why," Emerman told me. "We know at least that it is all a consequence of infections our ancestors had. So from there we want to try and derive a modern repertoire of antiviral genes."

They focused on chimpanzees, our closest relatives. Chimpanzees are easily infected by the AIDS virus, but it never makes them sick. That has remained one of the most frustrating mysteries of the epidemic. How did nearly identical genetic relatives become immune to a virus that attacks us with such vigor? The most dramatic difference between the chimp genome and ours is that chimps have roughly 130 copies of a virus called *Pan troglodytes* endogenous retrovirus, which scientists refer to by the acronym PtERV (pronounced "pea-terv"). Gorillas have 80 copies. Humans have none.

"We can see that PtERV infected gorillas and chimps four million years ago," Emerman told me. "But there was never any trace of its infecting humans." It is possible that all infected humans died, but it is far more likely that we developed a way to repel the virus. Nobody knew why until Emerman, Malik, and Shari Kaiser, a graduate student in Emerman's lab, presented evidence for a startling theory: the evolutionary process that protects us from PtERV may be the central reason we are vulnerable to HIV.

"We thought we must have a defense against this thing that they don't have," Malik told me, picking up the story the following day. Evolutionary biologists are not given to emotional outbursts — by definition, they take the long view. Malik is an engaging and voluble exception. When an antiviral protein excites him, he doesn't hold back. "Where but in evolutionary history can you see a story like this, with PtERV and the chimps?" he asked, leaping up from his chair to begin sketching viral particles on a whiteboard. "It's simply amazing."

He launched into a description of the complex interactions between viruses and the proteins that we have developed to fight them. There is one particular gene, called TRIM5α, which in humans manufactures a protein that binds to and destroys PtERV. "Our version of this gene is highly effective against PtERV, which is why we don't get infected," he said. Every primate has some version, but it works differently in each species — customized to fit the varying evolutionary requirements of each. In the rhesus monkey, that single gene provides complete protection against HIV infection. In humans it does nothing of the kind. "When Michael and I started to get into this business, people had never thought much about the evolutionary meaning of that gene. But we wondered, Is TRIM5α just an anti-HIV factor or is there something else going on here?"

Like the two human retroviruses that were reconstructed in France and in New York, PtERV has long been extinct; Emerman and Malik realized that they would have to assemble a new version if they hoped to learn how we became immune to it. They took scores of viral sequences and lined them up to see what they had in common. The answer was almost everything. When there were differences in the sequence, the researchers used a statistical model to predict the most likely original version. Then they put the virus back together. (Like Bieniasz in New York, they did so in such a way that the virus could reproduce only once.) They modified the human TRIM5α protein so that it would function like the chimp version. After that the protein no longer protected humans against the reconstructed copy of the virus. Next they tested this modified version against HIV. Emerman placed it in a dish, first with HIV and next with PtERV. What he found astonished him. No matter how many times he repeated the test, the results never varied. "In every case, the protein blocked either PtERV or HIV," Emerman told me. "But it never protected the cells from both viruses."

There are several possible ways to interpret the data, but the one favored by the researchers is that *because* humans developed an effective defense against one virus, PtERV, at about the time we split off from the chimps, 5 million years ago, we were left vulnerable to a new one, HIV. "If we can develop a drug that acts the same way the monkey version of this protein acts — so that it recognizes HIV and neutralizes it — we could have a very effective therapy," Malik said. Both he and Emerman stressed that this day will not come soon. "First, we have to establish what part of TRIM5α is actually responsible for protecting monkeys against HIV," Malik said. "Then we would have to try and make it as a drug" — one that the human body won't reject. "The challenge is to find out how little you can change the human version and still make it effective against HIV. That is really what drives this whole story of re-creating that extinct virus and doing these experiments. Nobody is doing this as a gimmick. This virus could open doors that have been closed to us for millions of years. And if we can learn how to do that we have a chance to find a very effective response to one of the world's most incredibly effective viruses."

The Oxford University zoology department is housed in a forbidding concrete structure that looks like an Eastern European police station. The building is named for the Dutch ethologist Niko Tinbergen, whose work with wasps and gulls, among other species, won him a Nobel Prize and helped establish the study of animal behavior as a science. Tinbergen's most famous student, Richard Dawkins, has carried on the university tradition of aggressive independence, and so have the younger members of the faculty. I stopped by the department a few months ago to have lunch with two of them, Aris Katzourakis and Robert Belshaw, both evolutionary biologists who have made the new field of paleovirology a specialty. Just before I arrived, Katzourakis had lobbed a bomb into the field.

Nobody knows what chain of evolutionary factors is required to transform an infectious virus — such as HIV — into one that is inherited. Such a virus would have to invade reproductive cells. HIV doesn't do that. It belongs to a class called lentiviruses (from the Latin for "slow"), which are common in mammals like sheep and goats. Because lentiviruses had never been found in any animal's genome, most virologists assumed that they had evolved recently. Until this summer, the oldest known lentivirus was "only" a million years, and almost no one thought that a lentivirus could become endogenous.

In a paper titled "Discovery and Analysis of the First Endogenous Lentivirus," published in the spring of 2007 in *Proceedings of the National Academy of Sciences,* Katzourakis, along with collaborators from Oxford, Stanford University, and Imperial College London, showed otherwise. They discovered the fossilized remains of an ancient lentivirus — the same type that causes AIDS — within the genome of the European rabbit *(Oryctolagus cuniculus).* "At first, I just assumed it was a mistake," Katzourakis told me over lunch in the building's cafeteria, Darwin's Café. "We checked it twice, three times. But we kept seeing genes that are found only in lentiviruses." They named their discovery "rabbit endogenous lentivirus type K," or RELIK. An obvious next step for Katzourakis and his group will be to work with virologists who can assemble a functional version of the ancient virus — as the researchers in Paris, New York, and Seattle have done. "It's the most promising way to explore the evolution and the impact of HIV," he said.

It might be more than that. AIDS researchers have always been handicapped by the absence of a small-animal model in which to study the effects of the disease. It is not easy to use monkeys or sheep. They are expensive and difficult to obtain, and for reasons of ethics, many experiments on them are proscribed. "Although RELIK is an ancient lentivirus and only defective copies were identified in this analysis," the authors wrote, "recent research has shown that it is possible to reconstruct infectious progenitors of such viruses," which, they concluded, could potentially "provide a small animal model for experimental research."

The discovery has already changed the way scientists think about viral evolution, and about HIV in particular. "The most obvious implication is that we can no longer say that HIV could not become endogenous," John Coffin, of Tufts, told me, though he still considers that unlikely. "It opens the field to a whole new level of examination." It also considerably alters the phylogenetic tree. RELIK is at least 7 million years old, which makes it the oldest known lentivirus. "It is possible that primate lentiviruses such as HIV and SIV" — its simian cousin — "are much older than people ever thought," Coffin said.

We can't be certain when endogenous retroviruses entered our genome, because it is impossible to watch a 5-million-year process unfold. Yet in Australia a retrovirus seems to be evolving in front of our eyes. Beginning in the late nineteenth century, koalas on the mainland were hunted nearly to extinction. To protect them, as many as possible were captured and moved to several islands in the south. In the past hundred years, those koalas have been used to replenish the population on the mainland and on several other Australian islands. In many cases, though, they have become infected with a retrovirus that causes leukemia, immune disorders, and other diseases. It can even kill them. The epidemic presents a significant threat to the future of the species, and scientists have followed it closely. One group, from the University of Queensland, looked for the virus in koala DNA — and, as one would expect with a retrovirus, found it. The team also noticed that some of the babies, known as joeys, were infected in the same locations on their DNA as their parents. That means that the virus has become endogenous. Yet when the scientists examined the koalas on Kangaroo Island, in the south, they discovered something they had not anticipated: none of the koalas were infected.

That could mean only one thing: since the infected animals had all been moved just in the past century, the koala retrovirus must have spread to Australia recently and is entering the genome now. That offers virologists and evolutionary biologists their first opportunity to learn how a virus transforms itself from something that can simply infect (and kill) its host to an organism that will become a permanent part of that host. Persistent viruses tend to grow weaker over the years. They couldn't live for long if they killed everything they infected. How they adapt, though, is a mystery. "Events like this have obviously occurred in human evolution," Paul Bieniasz told me — even with viruses like HIV. "We might be able to see how the koala infection settles into the genome and whether it plays a role in helping its host fend off other viruses," he continued. "Whatever we learn will be useful, because we could never have learned it in any other way."

In 1963 Linus Pauling, the twentieth century's most influential chemist, wrote an essay with Emile Zuckerkandl in which they predicted that it would one day become possible to reconstruct extinct forms of life. It has taken half a century for scientists to acquire the information necessary to master most of the essential molecular biology and genetics, but there can no longer be any doubt that Pauling was right. Once you are able to assemble the ancestral sequence of any form of life, all you have to do is put the genes together, and back it comes.

"The knowledge you gain from resurrecting something that has not been alive for a million years has to be immensely valuable," Harmit Malik told me in Seattle. "We didn't take it lightly, and I don't think any of our colleagues did, either." He repeatedly pointed out that each virus was assembled in such a way that it could reproduce only once. "If you can't apply the knowledge, you shouldn't do the experiment," he said. Malik is a basic research scientist. His work is not directly related to drug development or treating disease. Still, he thinks deeply about the link between what he does and the benefits such work might produce. That is an entirely new way to look at the purpose of scientific research, which in the past was always propelled by intellectual curiosity, not utilitarian goals. Among elite scientists, it was usually considered gauche to be obsessed with anything so tangible or immediate; brilliant discoveries were supposed to percolate. But that paradigm was constructed before laboratories around the world got into the business of reshaping, resurrecting, and creating various forms of life.

The insights provided by recent advances in evolutionary biology have already been put to use, particularly in efforts to stop the AIDS virus. One of the main reasons that endogenous retroviruses can enter our genome without killing us is that they make many errors when they reproduce. Those errors are genetic mutations. The faster a cell reproduces (and the older it is), the more errors it is likely to make. And the more errors it makes, the less likely it is to be dangerous to its host. "Viruses are accumulating and becoming more decrepit with every passing million years" was the way Malik described it to me. That realization has led AIDS researchers to contemplate a novel kind of drug. Until recently, antiviral medications had been designed largely to prevent HIV from reproducing. Various drugs try to interfere with enzymes and other proteins that are essential for the virus to copy itself. There is a problem with this approach, however. Because the virus changes so rapidly, after a while a drug designed to stop it can lose its effectiveness completely. (That is why people take cocktails of HIV medications; the combinations help slow the rate at which the virus learns to evade those interventions.)

Scientists at a company called Koronis Pharmaceuticals, just outside Seattle, are taking the opposite approach. They hope that by speeding up the life cycle of the AIDS virus they can drive it to extinction. The goal is to accelerate the virus's already rapid pace of mutation to the point where it produces such an enormous number of errors in its genome that it ceases to pose a threat. Like endogenous retroviruses, HIV would become extinct. Earlier this month, researchers at the University of California at San Francisco and at the University of Toronto announced an even more fascinating way to use the fossils in our genome. HIV infects immune-system cells and alters them so that they can produce more HIV. In doing so, they stimulate endogenous retroviruses, which then produce proteins that act as a sort of distress signal. Those signals can be detected on the surface of HIV-infected cells, and in theory it should be possible to develop vaccines that target them. In essence, such a vaccine would act like a smart bomb, homing in on a signal transmitted from within each HIV-infected cell. The team in San Francisco found strong evidence of those signals in the immune cells of fifteen of sixteen volunteers who were infected with HIV. In an uninfected control group, the signals were far weaker or were absent altogether. "For a vaccine against an infectious agent, this is a completely new strategy," Douglas Nixon, the immunologist who led the team, said. It's one that could not have emerged without the recent knowledge gained through experiments with endogenous retroviruses.

There may be no biological process more complicated than the relationships that viruses have with their hosts. Could it be that their persistence made it possible for humans to thrive? Luis P. Villarreal has posed that question many times, most notably in a 2004 essay, "Can Viruses Make Us Human?" Villarreal is the director of the Center for Virus Research at the University of California at Irvine. "This question will seem preposterous to most," his essay begins. "Viruses are molecular genetic parasites and are mostly recognized for their ability to induce disease." Yet he goes on to argue that they also represent "a major creative force" in our evolution, driving each infected cell to acquire new and increasingly complex molecular identities. Villarreal was among the first to propose that endogenous retroviruses played a crucial role in the development of the mammalian placenta. He goes further than that, though: "Clearly, we have been observing evolution only for a very short time. Yet we can witness what current viruses," such as HIV, "can and might do to the human population."

Villarreal predicts that without an effective AIDS vaccine, nearly the entire population of Africa will eventually perish. "We can also expect at least a few humans to survive," he wrote. They would be people who have been infected with HIV yet for some reason do not get sick. "These survivors would thus be left to repopulate the continent. However, the resulting human population would be distinct" from those whom HIV makes sick. These people would have acquired some combination of genes that confers resistance to HIV. There are already examples of specific mutations that seem to protect people against the virus. (For HIV to infect immune cells, for example, it must normally dock with a receptor that sits on the surface of those cells. There are people, though, whose genes instruct them to build defective receptors. Those with two copies of that defect, one from each parent, are resistant to HIV infection no matter how often they are exposed to the virus.) The process might take tens or even hundreds of thousands of years, but Darwinian selection would ultimately favor such mutations and provide the opportunity for the evolution of a fitter human population. "If this were to be the outcome," Villarreal wrote, "we would see a new species of human, marked by its newly acquired endogenous viruses." The difference between us and this new species would be much like the difference that we know exists between humans and chimpanzees.

For Villarreal and a growing number of like-minded scientists, the conclusion is clear. "Viruses may well be the unseen creator that most likely did contribute to making us human."

JEFFREY TOOBIN

The CSI Effect

FROM *The New Yorker*

ON THE EVENING OF MARCH 10, 2003, two New York Police Department detectives, James V. Nemorin and Rodney J. Andrews, were shot and killed in an unmarked police car while attempting an undercover purchase of a Tec-9 assault pistol on Staten Island. The case was significant not just because two officers had died but because the man who was eventually charged with the murders, Ronell Wilson, faced the possibility of becoming the first person in more than fifty years to be executed for a crime in New York State.

The government's theory was that Wilson, who was with an accomplice in the back seat of the car, shot the detectives during a robbery attempt. Among the evidence retrieved from the crime scene were hundreds of hairs and fibers, and prosecutors enlisted Lisa Faber, a criminalist and the supervisor of the NYPD crime lab's hair and fiber unit, to testify at Wilson's trial in the winter of 2006. Under questioning in Brooklyn federal court, Faber said that she had compared samples of fabric from the detectives' car with fibers found on gloves, jeans, and a baseball cap that Wilson had allegedly been wearing on the night of the crime. The prosecutor asked Faber to describe the methods and equipment she had used to make her analysis. Then she asked Faber what she had found.

"My conclusion is that all of those questioned fibers could have originated from the interior of the Nissan Maxima, from the seats, and/or the backrests," Faber said. She added that in her field "the strongest association you can say is that 'it could have come from'" the source in question.

Faber's testimony was careful and responsible — and not very significant. She could not say how common the automobile fabric that she had examined is or how many models and brands use it. Nor could she say how likely it was that the fabric from the car would show up on Wilson's clothes. Faber used no statistics, because there was no way to establish with any precision the probability that the fibers came from the detectives' car. DNA tests had proved that blood from one of the detectives was on Wilson's clothes, and based on this fact, as well as on testimony from his accomplice and from Faber, Wilson was convicted and sentenced to death. "Given how much evidence they had in the case, I wasn't crucial," Faber told me. "The prosecutors liked the idea of fiber evidence in addition to everything else. Maybe they thought the jury would like it because it was more *CSI*-esque."

CSI: Crime Scene Investigation, the CBS television series, and its two spinoffs — *CSI: Miami* and *CSI: New York* — routinely appear near the top of the Nielsen ratings. (A recent international survey, based on ratings from 2005, concluded that *CSI: Miami* was the most popular program in the world.) In large part because of the series' success, Faber's profession has acquired an air of glamour, and its practitioners an aura of infallibility. "I just met with the conference of Louisiana judges, and, when I asked if *CSI* had influenced their juries, every one of them raised their hands," Carol Henderson, the director of the National Clearinghouse for Science, Technology and the Law at Stetson University in Florida, told me. "People are riveted by the idea that science can solve crimes." At the Las Vegas criminalistics bureau, where the original version of the show is set, the number of job applications has increased dramatically in the past few years. In the pilot for the series, which was broadcast in 2000, Gil Grissom, the star criminalist, who is played by William Petersen, solved a murder by comparing toenail clippings. "If I can match the nail in the sneaker to the suspect's clipping . . ." Grissom mused, then did just that. In the next episode, the Las Vegas investigators solved a crime by comparing striation marks on bullets. "We got a match," one said. Later in the same show, Nick Stokes (George Eads) informs Grissom, his boss, "I just finished the carpet-swatch comparisons. Got a match."

The fictional criminalists speak with a certainty that their real-life counterparts do not. "We never use the word 'match,'" Faber, a thirty-eight-year-old Harvard graduate, told me. "The terminology is very important. On TV, they always like to say words like 'match,' but we say 'similar,' or 'could have come from' or 'is associated with.'"

Virtually all the forensic-science tests depicted on *CSI* — including analyses of bite marks, blood spatter, handwriting, firearm and tool marks, and voices, as well as of hair and fibers — rely on the judgments of individual experts and cannot easily be subjected to statistical verification. Many of the tests were developed by police departments more than a hundred years ago, and for decades they have been admitted as evidence in criminal trials to help bring about convictions. In the mid-1990s, nuclear-DNA analysis — which can link suspects to crime-scene evidence with mathematical certainty — became widely available, prompting some legal scholars to argue that older, less reliable tests, such as hair and fiber analysis, should no longer be allowed in court. In 1996 the authors of an exhaustive study of forensic hair comparisons published in the *Columbia Human Rights Law Review* concluded, "If the purveyors of this dubious science cannot do a better job of validating hair analysis than they have done so far, forensic hair comparison analysis should be excluded altogether from criminal trials."

Recently a commission on forensic science sponsored by the National Academy of Sciences held an open session in Washington at which several participants questioned the validity of hair and fiber evidence. Max Houck, the director of the Forensic Science Initiative at West Virginia University and the coauthor of an important study that reviewed hair analyses by the FBI, was fiercely criticized by several members of the commission, including one of the cochairs, Harry T. Edwards, a senior federal appeals court judge. "It sounds like there is a lot of impressionistic and subjective examination going on," Edwards said after Houck described the study. "Follow-up examiners repeated [the analyses] and made the same mistakes," Edwards said. "That's the scariest part."

Sir Robert Peel is credited with creating the first modern police force, the bobbies, in London in 1829, but the transformation of law enforcement, and especially forensic science, into a professional discipline was a haphazard affair. Scientists occasionally took an interest in police work, and courts sometimes accepted their testimony. Oddly, one prominent early figure in the field developed specialties in both bullets and hair, which have little in common except that both are often found at crime scenes. In 1910 Victor Balthazard, a professor of forensic medicine at the Sorbonne, published the first comprehensive study of hair, *Le Poil de l'Homme et des Animaux,* and three years later, in an influential article, he theorized that the grooves inside every gun barrel leave a unique imprint on bullets that pass through it. In the mid-1920s, Calvin Goddard, a New York doctor, began using a comparison microscope, which allows an analyst to examine two slides at the same time, to study bullets. In 1929 he analyzed bullets collected at the site of the St. Valentine's Day massacre, in which gunmen wearing police uniforms shot and killed six members of George (Bugs) Moran's gang and a seventh man. Goddard test-fired all eight machine guns owned by the Chicago police and found no match with the bullets used in the crime. Two years later, he examined two submachine guns retrieved from the home of Fred Burke, a sometime hit man for Al Capone, Moran's great rival. Goddard pronounced Burke's guns the murder weapons, and the feat so impressed local leaders that they established a crime lab, the nation's first, and installed Goddard as its director.

The comparison microscopes in the NYPD crime lab are more powerful than Goddard's model was, but many of the techniques the lab uses haven't changed substantially in decades. Situated in a remote part of Queens, in a sprawling office building that formerly belonged to York College of the City University of New York, the crime lab combines municipal decrepitude and state-of-the-art technology. The seating area in the lobby includes an old minivan bench, complete with a dangling seat belt. There are six analysts in Faber's hair and fiber unit, and each has a polarized-light microscope, which costs about $15,000. In addition, the unit has two comparison microscopes, which cost $50,000, but only one phone line, which doesn't have voicemail. Having worked for the NYPD for nearly a decade, Faber has acquired a weary proficiency in the department's eccentricities. "We have some of the best lab equipment in the country, maybe as good as the FBI," she told me recently, when I visited her at the lab. "But for the little stuff we have to scrounge."

Faber's father, a German businessman, was transferred by his company to Manchester, New Hampshire, in 1966, two years before Lisa was born. She was in junior high school when she became interested in trace-evidence analysis, which includes hair and fiber. "It was the time of the Atlanta child murders," she recalled, referring to the killings of more than two dozen children and young men in the city between 1979 and 1981. "I would sit there watching the news and see how they were connecting the murders through fibers at the murder scenes. It just fascinated me that you could have a body floating in a river and still find fiber evidence that would connect it to a car." The trace evidence was critical in the case against Wayne Williams, a twenty-three-year-old aspiring music promoter, who was convicted of two of the murders and sentenced to two life terms. No one has been charged in connection with the other deaths. (In February 2007, following years of unsuccessful appeals by Williams, lawyers for Georgia agreed to allow DNA tests on some of the hair used as evidence in the case. Results of the tests have not yet been reported.)

At Harvard, Faber majored in East Asian studies. "I still followed murder cases, and the Internet was just getting started not long after I graduated," she recalled. "So I looked around to see what kinds of graduate programs in forensic science there might be around the country. There were four." One was at George Washington University, where she enrolled. Her first class, in trace analysis, was taught by a former FBI special agent named Hal Deadman. "He passed around materials on that first day, and they were from the Atlanta murders. He had been an investigator on the case. I thought, I can't believe this. This is perfect for me." Two years later, having completed a master's degree, Faber joined the NYPD, where she earns about $70,000 a year. A tall woman who usually wears her long blond hair in a barrette, Faber has an apartment in Greenwich Village, where few other NYPD employees live. "No one ever steals my lunch out of the office refrigerator," she said. "People ask me, 'That's all you're eating — vegetables?'"

When a piece of evidence, usually a garment, arrives in Faber's unit, a junior analyst pats a strip of clear tape on the fabric to pick up any loose hairs or fibers. The tape strips are then placed face down on a plastic sheet, which is given to a more senior person to study under a relatively low-powered stereo microscope. "I've got to do a screening at this point just to see what might be useful to examine in greater detail," Faber told me. "As far as hair goes, I'm looking only for head hair and pubic hair, because arm or chest hair doesn't have enough variation to be useful for comparison. If I see a fiber, I'll circle it with my green Sharpie. If I see a hair, I'll circle it in red. A brown circle means I'm not sure what it is. When I'm done, I take my tweezers and remove each hair and fiber and glue them onto slides. Now I'm really ready to begin my analysis."

According to a long-established practice of hair analysis, the examiners study between twenty and thirty characteristics of each hair. (Only hairs whose roots are intact — typically because they have been pulled from someone's head — and are in the growing stage have nuclear DNA, which is unique to each person; the vast majority of hairs found at crime scenes lack roots suitable for testing.) "The first thing we look at is pigment — color," Faber said. "Is there dye or bleach, and how much of it? We like it when there's dye, because it makes a hair distinctive. Then there is the cuticle, which is like the yellow on a pencil — the outside — and we see if it's damaged or serrated." One of the things that make hair analysis so challenging is that a person can have hairs of many different colors on his or her head. Nor can ethnicity be established with certainty on the basis of hair samples alone. "Hair is curly because of diameter variation," Faber explained. "A black person's hair typically has a lot of diameter variation. Asian people have little variation, which is what makes it straight." Faber has hundreds of slides of hair that she has collected from around the world and uses to train junior analysts. "I ask my friends to give me their hair," she said.

For the Wilson case, Faber examined more than a hundred pieces of evidence, including the victims' clothes, the suspects' clothes, and the interior of the car, along with two do-rags and other items seized from the street. From this material, Faber recovered thousands of fibers and hundreds of hairs, each of which had to be assessed under a microscope.

"The examinations involving the car were especially difficult and time-consuming," she said. For Faber, the key question in the case was whether she could identify fibers on Wilson's clothing that resembled those taken from the car. "There were hundreds of fibers that looked stereoscopically similar at lower-power magnifications — that is, up to fifty times — but many of them were in fact different in chemical composition," she said. "I could tell them apart only when I mounted each fiber on a slide and examined it under two hundred to four hundred magnification with a polarized-light microscope. It's only at that level that you can see what you need to know to identify the chemical composition of the fiber, like whether it's polyester, nylon, or acrylic." Once she had eliminated those fibers that could not have come from the car, Faber performed two additional tests on the remaining set: Fourier transform infrared (FTIR) spectroscopy, to establish and compare chemical composition, and microspectrophotometry (MSP), to compare color. Working quickly, Faber said, she could complete perhaps five FTIR and MSP analyses a day; evaluating all the evidence in the case took her two and a half years.

Several times a year, at the New York Police Academy, Faber lectures about trace analysis in a criminal investigations course, which generally attracts about a hundred midcareer detectives and other investigators. Earlier this spring, she had the misfortune of appearing late in the three-week course, which meant that the students — about ninety beefy men and six women in an overheated classroom — listened and observed in varied states of consciousness.

"With trace analysis, what we are doing is comparing a q to a k," Faber told the group. "A q is a questioned sample from a crime scene — a paint chip on someone's clothing from a hit and run, a hair in the hand of a murder victim. The source is unknown. The k is a known sample — a hair from the autopsy, or one that you take from a suspect. What we do in the trace-analysis unit are comparisons with q's and k's. We see if there is a connection."

Faber's brief summary defined the dilemma at the heart of forensic science. "There are really two kinds of forensic science," says Michael J. Saks, a professor of law and psychology at Arizona State University and a prominent critic of the way such science is used in courtrooms. "The first is very straightforward. It says, 'We have a dead body. Let's see what chemicals are in the blood. Is there alcohol? Cocaine?' That is real science applied to a forensics problem. The other half of forensic science has been invented by and for police departments, and that includes fingerprints, handwriting, tool marks, tire marks, hair, and fiber. All of those essentially share one belief, which is that there are no two specimens that are alike except those from the same source." Saks and other experts argue that there is no objective basis for making the link between a q and a k. "There is no scientific evidence, no validation studies, or anything else that scientists usually demand, for that proposition — that, say, two hairs that look alike came from the same person," Saks said. "It's the individualization fallacy, and it's not real science. It's faith-based science."

Virtually all experts agree about the reliability of DNA evidence. "DNA is based on a well-known technology and scientific principles that have a lot of uses outside the lab and a lot of good validation data," D. Michael Risinger, a professor at Seton Hall University School of Law, said. "You will typically know what the error rates are. The tests produce actual probability statements about results. It's real science." Currently, two kinds of DNA are subject to forensic testing: nuclear DNA and mitochondrial DNA (mtDNA), which is passed from mother to child. Unlike nuclear DNA, mtDNA can be extracted even from hairs that lack roots; though mtDNA tests cannot establish a precise match, they can eliminate many potential suspects. In the past two years, New York authorities have begun conducting mtDNA tests on some evidence.

At the recent session on forensic science in Washington, Max Houck explained that he and his coauthor, Bruce Budowle, a senior scientist at the FBI Laboratory, had used mtDNA technology to test the validity of hair analysis. The authors reviewed the results of 170 microscopic hair examinations, which produced 80 associations between q and k samples. But subsequent mtDNA tests of the hairs showed that in 9 cases — more than 10 percent — the samples could not have come from the same person. The number of errors was concerning, Judge Edwards said. According to his calculations, he added, the study's error rate was actually close to 35 percent. How, he asked Houck, was such a flawed process acceptable?

Houck challenged Judge Edwards's use of the word "error," arguing that mtDNA tests provided a way to refine the more general conclusions of microscopic examinations. To illustrate his point, Houck offered an analogy: "Suppose you have an art expert who is looking at three possible van Goghs. He sees that all three are consistent with his style. But then a chemical test comes along that shows two of them were painted after van Gogh died. They're not van Goghs, but that doesn't mean that the art expert was wrong about them. They may have been consistent with his style. It's the same with hair analysis. The subsequent DNA tests don't mean that the original tests were wrong, just that a more refined test has come along."

"I don't think your analogy holds at all," Edwards said. "Those are real errors, and this is not about art history. My great worry is that there are a lot of people going to jail on bad information."

Margaret A. Berger, a professor at Brooklyn Law School, added, "We know from a lot of DNA exonerations that they come from bad hair evidence. So the real question raised by your study is whether the courts should ever allow microscopic evidence when there is no DNA to back it up? If there is no mtDNA, I think it should be excluded."

"Let me ask you this," Channing R. Robertson, a professor of chemical engineering at Stanford, said to Houck. "Suppose your son or daughter was accused of a crime, and someone came on the stand and gave their qualifications as a hair examiner and made an association based on microscopic examination, and that led to the conviction of your child. Would you feel that justice had been served?"

After an awkward pause, Houck said, "Not unless there was mtDNA as well."

Houck's critics focused on the possibility of errors by well-meaning hair and fiber analysts, but the field has also been beset by scandals. Incompetent or malevolent trace-evidence examiners in several states, including Texas, West Virginia, and Illinois, have produced scores of tainted verdicts, many of which have been uncovered by the Innocence Project, the legal advocacy group founded fifteen years ago by the defense attorneys Barry Scheck and Peter Neufeld. In a 1987 case involving the rape of an eight-year-old girl, Arnold Melnikoff, the manager of the Montana state crime lab, testified that there was a "less than one in ten thousand chance" that hairs found at the crime scene did not belong to the defendant, Jimmy Ray Bromgard, who was convicted. In 2002 a DNA test of sperm found on the victim's underwear established Bromgard's innocence, and several more cases in which Melnikoff testified have been overturned or called into question. (John Grisham's recent bestseller, *The Innocent Man,* is a nonfiction account of a 1988 murder case in Oklahoma in which faulty hair analysis, among other things, led to an unjust conviction.) As Faber acknowledges, "There have been horrible cases involving bad hair examiners." (In 2007 the NYPD released the results of an internal investigation of its crime lab, which revealed that in 2002 two technicians in the controlled-substances division had failed a department proficiency test. This information was not reported to the national accreditation body for forensic labs, as the department is required by law to do. The police commissioner, Raymond W. Kelly, replaced the deputy chief who supervised the lab and ordered the creation of an oversight panel.)

Faber and her colleagues, unlike their fictional counterparts on TV, do not visit crime scenes, leaving evidence collection to specially trained NYPD officers. Nevertheless, working in the hair and fiber lab has given Faber a macabre expertise in city life. She is often called on to examine the clothing of crime victims and suspects, and she has noticed that patterns recur. "They like big clothes," she told me, referring to the suspects. "We frequently see sizes like 4X."

Faber has recently taken steps to combine traditional hair and fiber analysis with more modern and accurate technology. One day last month, Faber's colleague Debbie Hartmann was working over a gray Rocawear sweatshirt with tape and tweezers, trying to find hairs to analyze. The sweatshirt had been left behind during a burglary of a house in Queens, and Hartmann was examining it as part of a new project that Faber has implemented in the crime lab. "The city has had a program in place for a long time where we see if we can do DNA testing in every murder and rape," Faber told me. "But we thought there was no reason not to do burglaries as well. We know that burglars often change their clothes when they're inside a house so they don't look the same coming out as they did coming in. They leave their tools, like screwdrivers, behind. They eat food and drink the beer in the fridge. They don't want to leave fingerprints, so they wear rubber gloves and leave them behind. That turns out to be better, because the gloves have DNA from their skin cells, which is better evidence than fingerprints. All of that stuff they leave behind can be tested for DNA, so we decided to do it." (The NYPD crime lab, unlike police departments in many other cities, does no DNA testing itself; Faber and her colleagues send evidence for testing to the medical examiner in the Department of Health or to private contractors.)

The burglary program, which is known as Biotracks and is funded with a grant from the National Institute of Justice, began in 2003 with a pilot program in northern Queens. It was soon expanded to cover the whole city. As of last week, Faber's team — it consists of, in addition to Faber, a police sergeant and another criminalist — had analyzed 2,330 cases and found DNA profiles in 1,541 of them. The results of these tests have been placed in the FBI's Combined DNA Index System (CODIS), a database of more than 4 million DNA profiles from across the country, and almost 300 suspects have been identified.

Some of Biotracks' success stories resemble plots of *CSI* episodes. In 2004 Stanley Jenkins, a forty-seven-year-old man who had a criminal record, broke into a residence in Queens and stole jewelry and camera equipment. He used a tissue to wipe away his fingerprints, and no prints were found in the house. But the tissue yielded a DNA sample, and Jenkins was convicted after a trial and sentenced to twenty years to life. The same year, Faber's team found matching DNA on broken windows in a series of smash-and-grab burglaries at several Manhattan stores, including Tourneau, Coach, and Fendi. Police officers picked up a man named David Gibson because he resembled the suspect in the case, and detectives noticed that he had a bandaged wound on his left hand. They offered to change the dressing and submitted the old bandage for DNA testing, which tied Gibson to the crimes.

In 2005, in perhaps the most bizarre case to date, the Biotracks group took a DNA sample from a soda bottle that a burglar had drunk from at a house in the Bronx. The DNA profile identified a pair of identical twins, Kenneth Williams and Andre Fuller, who were both in the CODIS system because they had criminal records. (Only identical twins have the same nuclear DNA.) It turned out that at the time of the burglary Williams was incarcerated, so when Fuller was arrested in an unrelated case he was charged with the burglary. (The case is pending.) "We never did figure out how twins had different last names," Faber said.

Faber regards Biotracks as one way of tethering her field to the rigorous science of DNA. "I know everyone wants to go straight to the DNA lab these days, and I don't blame them," she said. "But they have to come to us first, and for good reason. Ninety percent of the hair recovered at a crime scene doesn't have roots in the right stage of growth. You need roots for nuclear DNA. You might ask, Why is microscopic trace examination necessary? Why not submit all hairs for DNA analysis immediately? But you cannot go to the medical examiner with two hundred hairs. We have so many crimes and hairs. Hairs are everywhere. We couldn't possibly give every hair to the m.e. We have to screen it first, see if there is a possibility of any kind of DNA test, and then we can send it over."

There is wisdom, but also some poignancy, in Faber's efforts to modernize the NYPD's hair and fiber unit. The success of the Biotracks program suggests that the best thing hair analysts can do is turn themselves into initial screeners for the test that really matters: DNA. Soon the demand for the kind of testimony that Faber provided in the Ronell Wilson case may be limited, especially if the commission sponsored by the National Academy of Sciences publishes a report harshly critical of the field's current practices. (In 1992 a similar commission released a blue-ribbon federal report that resulted in the widespread use of DNA evidence in courtrooms.)

Faber is ambivalent about *CSI* — flattered by the attention it has brought to her profession but mindful of the cinematic license taken by the program's creators. She has noticed, for example, a new kind of applicant for jobs at the crime lab. "They're waiting for their interviews, and they look like they're auditioning for a hip profession," she said. "It's not the nerdy-looking people anymore. They don't realize that there is nothing cool or funky about this job."

A few years ago, when CBS was considering launching the *CSI: New York* spinoff, Faber showed Anthony Zuiker, the creator of *CSI,* and his colleagues around the crime lab. "They were great, really nice, and we were happy that they decided to do the show," she said. "But in the end they decided to do it in their studio in California."

ANDREAS VON BUBNOFF

Numbers Can Lie

FROM *The Los Angeles Times*

SAGITTARIANS are 38 percent more likely to break a leg than people of other star signs — and Leos are 15 percent more likely to suffer from internal bleeding. So says a 2006 Canadian study that looked at the reasons residents of Ontario had unplanned stays in the hospital.

Leos, Sagittarians: there's no need to worry. Even the study's authors don't believe their results. They're illustrating a point: that a scientific approach used in many human studies often leads to findings that are flat-out wrong.

Such studies make headlines every day, and often, as the public knows too well, they contradict each other. One week we hear that pets are good for your health, the next week that they aren't. One month cell-phone use causes brain cancer; the next month it doesn't.

"It's the cure of the week or the killer of the week, the danger of the week," says Barry Kramer, associate director for disease prevention at the National Institutes of Health in Bethesda, Maryland. It's like treating people to an endless regimen of whiplash, he says.

Take the case of just one item: coffee. Drinking two or three cups per day can triple the risk of pancreatic cancer, according to a 1981 study. Not so, concluded a larger follow-up study published in 2001. Coffee reduces the risk of colorectal cancer, found a 1998 study. Not so, according to one published later, in 2005.

"I've seen so many contradictory studies with coffee that I've come to ignore them all," says Donald Berry, chair of the department of biostatistics at the University of Texas M. D. Anderson Cancer Center in Houston.

"What about the man on the street?" asks Stan Young, a statistician at the National Institute of Statistical Sciences in Research Triangle Park, North Carolina. "He reads about coffee causing and not causing cancer — so many contradictory findings he begins to think, 'I don't trust anything these scientists are saying.'"

These critics say the reason this keeps happening is simple: far too many of these epidemiological studies — studies in which the habits and other factors of large populations of people are tracked, sometimes for years — are wrong and should be ignored. In fact, some of these critics say, more than half of all epidemiological studies are incorrect.

The studies can be influential. Often, in response to them, members of the public will go out and dose themselves with this vitamin or that foodstuff. And the studies also influence medical practice; doctors, the critics note, encouraged women to take hormones after menopause long before their effects were tested in randomized clinical trials, the gold standard of medical research.

Some of epidemiology's critics are calling for stricter standards before such studies get reported in medical journals or in the popular press. Young, one of the foremost critics, argues that epidemiological studies are so often wrong that they are coming close to being worthless. "We spend a lot of money and we could make claims just as valid as a random number generator," he says.

Epidemiology's defenders say such criticisms are hugely overblown. They are "quite simplistic and exaggerated," says Meir Stampfer, a professor of epidemiology and nutrition at the Harvard School of Public Health and a professor of medicine at Harvard Medical School.

What's more, some things simply cannot be tested in randomized clinical trials. In certain cases, to do so would be unethical. (Care to assign half the people in a trial to smoke cigarettes?) In other cases, a trial of adequate size and duration — say, to test whether coffee drinking raises or lowers the risk of Parkinson's disease — would have to control the habits of huge numbers of people for decades. That would be not only hugely expensive but also virtually impossible.

Stampfer cites examples of findings of epidemiology that, he says, have stood the test of time: smoking's link to lung cancer, to name the most notable.

Watching for Clues

In epidemiological studies (also called observational studies), scientists observe what's going on; they don't try to change it. From what they observe, they reach conclusions — for example, about the risk of developing a certain disease from being exposed to something in the environment, a lifestyle, or a health intervention.

There are different ways to do this. Cohort studies follow a healthy group of people (with different intakes of, say, coffee) over time and look at who gets a disease. These are considered the strongest type of epidemiological study.

Case-control or retrospective studies examine people with and without a certain disease and compare their prior life — for how much coffee they drank, for example — and see if people who got the disease drank more coffee in their past than those who didn't. Cross-sectional studies compare people's present lifestyle (how much coffee they drink now) with their present health status.

Epidemiological studies have several advantages: they are relatively inexpensive, and they can ethically be done for exposures to factors such as alcohol that are considered harmful, because the people under study chose their exposure themselves.

But epidemiological studies have their minuses, too, some of which are very well known. Suppose a study finds that coffee drinkers are more likely to get a certain disease. That doesn't mean coffee *caused* the disease. Other, perhaps unknown, factors (called "confounders" in the trade) that are unrelated to the coffee may cause it — and if coffee drinkers are more likely to do this other thing, coffee may appear, incorrectly, to be the smoking gun.

A much clearer picture of the role of coffee in disease could be found, in theory, via a randomized clinical trial. Such a study would divide a population in two, put one group on coffee and the other not, then follow both groups for years or decades to see which group got certain diseases and which didn't. The problem, however, is that a randomized trial is very expensive and takes a long time, and it can be difficult to control people's lives for that length of time.

Despite their shortcomings, epidemiological studies are often taken seriously, so much so that they can change medical practice. Such was the case after dozens of epidemiological studies, including one large, frequently cited study that came out of Harvard in 1991, had shown that taking estrogen after menopause reduced women's risk of getting cardiovascular disease.

"There was such a belief," even within the medical community, that hormone replacement became part of standard medical practice, says Lisa Schwartz, an associate professor of medicine at Dartmouth Medical School in Hanover, New Hampshire, even in the face of a potential increased risk of breast cancer. In fact, some scientists and doctors said it would be unethical to do a randomized clinical trial to check if the hormone effect was real.

But in the hormone epidemiological studies, women choosing to take hormones may well have been healthier in other ways, Kramer says. And the fact that they were healthier could explain the lower risk of heart disease, not the hormones.

"To get hormone therapy, you have to go to a doctor and have to have insurance," Kramer says. "That means you are in the upper strata of society."

Eventually, a randomized clinical trial was conducted as part of the Women's Health Initiative. Findings published in 2002 not only found no protection to the heart from hormone therapy but actually reported some harm.

Epidemiology's detractors say they have no trouble finding cases other than hormones where frequently cited and sometimes influential epidemiology studies have later turned out to be wrong or exaggerated.

In 1993 Harvard University scientists published two cohort studies reporting that vitamin E protected people from coronary heart disease. One, the Nurses' Health Study, followed more than 87,000 middle-aged female nurses without heart disease for up to eight years. It found that the 20 percent of nurses with the highest vitamin E intake had a 34 percent lower risk of major coronary disease than those with the lowest 20 percent of intake.

The other study followed almost 40,000 male health professionals without heart disease for four years and found a 36 percent lower risk of coronary disease in those men taking more than 60 international units (IU) of vitamin E per day compared with those consuming less than 7.5 IU.

In the three years after these studies appeared, each was cited by other research papers more than four hundred times, according to

John Ioannidis, an epidemiologist at the University of Ioannina School of Medicine in Ioannina, Greece. Vitamin E therapy for heart patients became widespread; a 1997 survey published in the *American Journal of Cardiology* reported that 44 percent of U.S. cardiologists routinely used antioxidants, primarily vitamin E.

The therapy was finally put to the test in a Canadian randomized clinical trial of about 2,500 women and 7,000 men aged fifty-five years or older who were at high risk for cardiovascular events. The findings, reported in 2000, showed that an average daily dose of 400 IU of vitamin E from natural sources for about four and a half years had no effect on cardiovascular disease.

Yet, Schwartz says, seven years after that finding, her patients continue to take vitamin E in the belief that it will protect their hearts. "I am still taking people off vitamin E," she says of her patients, some of whom have heart disease.

Study of Studies

In a provocative 2005 paper, Ioannidis examined the six most frequently cited epidemiological studies published from leading medical journals between 1990 and 2003. He found that in four of the six studies, the findings were later overturned by clinical trials. Vitamin E didn't protect the heart in men or women. Hormone therapy didn't protect the heart in women. Nitric oxide inhalation didn't help patients with respiratory distress syndrome. Another finding turned out later to be exaggerated: taking flavonoids reduced coronary artery disease risk only by 20 percent, not by 68 percent, as originally reported.

The only finding of the six that stood the test of time was a small study that reported that a chemical called all-trans retinoic acid was effective in treating acute promyelocytic leukemia.

The studies that overturned each of these epidemiological findings, Ioannidis says, "caused major waves of surprise when they first appeared, because everybody had believed the observational studies. And then the randomized trials found something completely different."

To be fair, Ioannidis also tested whether the most frequently cited *randomized* studies held up. He found that these had a much better track record. Only nine of thirty-nine oft-cited studies were later contradicted or turned out to be exaggerated when other randomized studies were done. True, Ioannidis looked at only six studies. But Young says he sees the same trend in his own informal counts of epidemiological claims. "When, in multiple papers, fifteen out of sixteen claims don't replicate, there is a problem," he says.

Belief can be costly, Young adds. For example, one part of the large, randomized Women's Heath Initiative study tested the widely held belief — based in large part on epidemiological studies — that a low-fat diet decreases the risk of colorectal cancer, heart disease, or stroke. The findings suggested that there was no effect; "$415 million later, none of the claims were supported," Young says.

Other scientists, while more cautious than epidemiology's most outspoken detractors, agree that there are many flawed studies. When Kramer first saw Ioannidis's number, "I said to myself, 'It can't be that bad,'" he says. "But I can't prove that it isn't. I know there are a lot of bad studies out there."

Ioannidis, Kramer says, is voicing what many know to be true.

Method in Doubt

Why does this happen?

Young believes there's something fundamentally wrong with the method of observational studies — something that goes way beyond that thorny little issue of confounding factors. It's about another habit of epidemiology that some call data mining.

Most epidemiological studies, according to Young, don't account for the fact that they often check many different things in one study. "They think it is fine to ask many questions of the same data set," Young says. And the more things you check, the more likely it becomes that you'll find something that's statistically significant — just by chance or luck, nothing more.

It's like rolling a pair of dice many times without anyone looking until you get snake eyes and then claiming you'd only rolled it once. Often epidemiological researchers ask dozens, maybe hundreds, of questions in the questionnaires they send to the people they study. They ask so many questions that something eventually is bound to come out positive.

That's where the Canadian star-sign study comes into play, Young says. It was only because the authors deliberately asked a lot of questions — to prove a point — that it was able to come up with significant results for something that couldn't be true.

The study's lead author, statistician Peter Austin of the Institute for Clinical Evaluative Sciences in Toronto, says that once he cleaned up his methodology (by adding a statistical correction for the large number of questions he asked) the association between Leos and internal bleeding and Sagittarians and leg breaks disappeared.

On the Defensive

Many epidemiologists do not agree with the critics' assertion that most epidemiological studies are wrong and that randomized studies are more reliable.

"Randomized studies often contradict one another, as do observational studies," says Harvard's Stampfer, who is an author on both of the frequently cited vitamin E and hormone replacement studies that Ioannidis says were later refuted.

Instead, Stampfer says, the two types of studies often test different things. "It's not an issue here that observational studies got it wrong and randomized trials got it right," he says, referring to the hormone replacement studies. "My view is that [both] were right and they were addressing different questions."

For one thing, the randomized studies on hormone replacement and vitamin E that Ioannidis cited in his 2005 paper looked at different populations than the observational studies they refuted, says Stampfer, who takes vitamin E himself.

In the hormone replacement case, the observational studies looked at women around the age of menopause. The randomized trial looked at women who were mostly well past that age.

In fact, Stampfer says, a recent reanalysis of the Women's Health Initiative data suggested a trend that hormone therapy may be less risky for younger than for older women. The effect was not statistically significant, but, Stampfer says, it's further support for the idea that hormones have different effects depending on when women start taking them.

And the vitamin E studies? The 1993 observational studies followed people who didn't have heart disease. The randomized study looked at people with known heart disease who were on many other medications. All those meds could easily override the effect of vitamin E, says Walter Willett, a professor of epidemiology and nutrition at the Harvard School of Public Health, who was a coauthor on both the hormone and the vitamin E epidemiological studies.

And, finally, the low-fat trial from the Women's Health Initiative. It's not surprising, Willett and Stampfer say, that this gold-standard trial failed to find what epidemiology had: that low-fat diets ward off heart disease, colorectal cancer, and stroke; the women in these trials didn't stick to their diets.

"The compliance with the low-fat diet was definitely far lower than anticipated," Willett says, "and probably far worse than even acknowledged in the papers."

Such arguments do not sway epidemiology's detractors. Each time a study doesn't replicate, "they make a specific argument why the studies are different," Young says. He concedes that epidemiology did uncover the truth about the risks of smoking — but only because the effects are so strong. "Even a blind hog occasionally finds an acorn," he says.

Yet epidemiologists warn that discarding results because of a correction for multiple testing may risk missing true and important effects — especially in cases where there's a good biological reason suggesting an effect, such as in studies of drugs that have been shown to work in animal experiments. And setting the bar too high might sometimes be dangerous, says Sander Greenland, a professor of epidemiology and statistics at UCLA. "Do you want to screen for medical side effects with the attitude, 'So what if we miss side effects?'" he asks. "That's deadly. That's ridiculous!"

The debate is unlikely to be resolved any time soon. "If you put five epidemiologists and five statisticians in a room and have this debate," Young says, "and try to get each one to convince the other side, at the end of the day it will still be five to five."

FLORENCE WILLIAMS

A Mighty Wind

FROM *Outside*

FROM THE TOP of a 164-foot-tall metal shaft perched on the small
Danish island of Samsø, the wind can just about suck the eyelashes
off your face.

After climbing 140 ladder rungs, I'm gripping the railing inside
the vibrating engine room, a bread-truck-size space at the tippy-top
of a windmill that holds a series of gyrating generators. At the press
of a button, the ceiling peels away, James Bond style, to reveal a
cloudless sky, interrupted only by the regular rotations of three
sleek, white, 88-foot-long blades.

The humming platform feels and sounds like a jet about to lift
off, but all the kinetic energy ends up beside me in the ten-ton
gearbox — there and in my hair. It's exhilarating, like riding a
Harley through a hurricane. I'm almost tempted to burst into a
Springsteen song. What I'm feeling at the base of my spine is the
prevalent weather pattern roiling in from the North and Baltic
seas: 20-mile-per-hour winds being converted into 463 kilowatts of
electricity, enough to power more than 600 households this very
moment.

And the view's nice, too.

I see the expanses of productive corn and pumpkin fields, the
Ben & Jerry–esque cows munching local silage among the sunflowers, the bike lanes, the sensible traffic flow, the traditional
timber-frame and brick houses, including the one belonging to the
proud farmer who owns this turbine. It's enough to make a New
Urbanist weep. From up here, Samsø — a dollop of land some
twelve miles off Denmark's main peninsula — looks like a marvel

of social and natural engineering. To the south of me stand two more windmills; to the northeast, eight more, like pins on a military map, and about two miles offshore is a necklace of an additional ten.

The land turbines provide enough electricity for the entire Nantucket-size island of about 4,200 people. The clean power generated by the offshore turbines is just gravy; sold to the mainland, it more than atones for the amount of carbon dioxide that Samsø pumps into the air through its use of "dirty energy" like petroleum, which fuels its cars and some home furnaces. In fact, Samsø has spent the past decade becoming an eco-wonderland, setting up wind, solar, biofuel, and other renewable technologies to satisfy its energy needs. The island has even gone beyond "carbon neutrality," the cherished environmental goal of zeroing out the production of CO_2, the greenhouse gas most responsible for global warming.

Now, in addition to its renown as a land of new potatoes, Samsø enjoys the distinction of being the most carbon-negative settlement of its size on Earth.

It's a relief to be on the Green Isle, a CO_2-safe zone where for once my everyday habits won't add to climate change. I've come to Denmark determined to be an ultra-low-carbon traveler, but I spent my first few days in Copenhagen, where I really had to work at it. I rode a Dahon folding bike all over the place and calculated the CO_2 output of every mile of public transport so I'd know how many trees to plant or clean-energy credits to buy when I got back home to make up for my carbon transgressions. In the biggest sacrifice, I spent three nights sleeping in one room with thirty-three other people at the Sleep-in Green hostel in Copenhagen. I took to calling it the Can't-Sleep-in Green. So what if the bunk beds squeaked and the girls from Texas donned their boots at five in the morning (and left large canisters of hair products that simply would not sort into any of the dazzling array of recycling bins)? It was great for my planetary balance sheet: solar electricity, organic juices, energy-efficient light bulbs, and minimal waste.

To get with the program even more, I sought tips from a local carbon guru, Bente Hessellund Andersen, who's known for her polite climate harangues directed at nongreen European leaders. A longtime environmentalist, she's neat and pretty, a soccer mom without the SUV. She doesn't even own a car. "I feel sad when young people tell me they need to have a car — very sad — since it's their kids who'll suffer from global warming," she told me during a torrential rainstorm at a bar near Tivoli Gardens. We were both sopping wet, having arrived, naturally, by bicycle.

Each Dane produces on average thirteen tons of carbon dioxide per year, Andersen said. (She herself produces less than that.) Each American produces about twenty. "We call it 'the American lifestyle' when we eat a lot of meat," she said, and proceeded to explain the golden rules of a low-carbon life, ticking off a list that made McDonald's look like the seventh ring of hell: eat vegetarian, locally grown foods, preferably organic, because they require less gasoline to transport and fewer fossil-fuel-laden pesticides and fertilizers. If you're really feeling zealous, eat your food cold, or, if you need it hot, keep a lid on the pot. "And as a tourist, buy locally made goods," she added. I don't think she had in mind my latest souvenir: a gel-filled plastic pen with a floating Viking ship inside.

But the biggest carbon spewers, Andersen went on, are transportation and housing. Together they contribute 60 to 80 percent of a person's CO_2 footprint. I was off the hook with my carrot-slaw consumption and with the hostel and the bike, but the planes and trains I'd taken had done some damage. Andersen referred me to some Web-based carbon calculators to figure out my transportation debt and told me to consult a report about food-associated CO_2 emissions, appetizingly called "Eating Oil." It might help me assess how many greenhouse gases were released to make my twelve-dollar lime-drenched mojito the other night at a restaurant called Pussy Galore's Flying Circus. I should have had an aquavit.

On Samsø, though, I could stop tallying and relax. Some islands feel magical because of healing waters or succulent fruits. Samsø's aura comes from the fetching way the islanders crunch their numbers to achieve the great Kyoto-driven concept known as offsets. Here's how it works: if you use dirty power, which most of us must, you just have to make up for it. Islands, nations, or citizens can take steps like planting thousands of trees to cancel out the carbon dioxide they've added to the atmosphere. They can also produce clean energy for others to use, thereby preventing additional emissions of greenhouse gases. With enough of these measures, it's possible — on paper, anyway — to become carbon neutral or even carbon negative. Samsø's wind turbines, for example, produce about 105 million kilowatt-hours of electricity per year, but the island needs only a quarter of that, leaving a surplus of 77 million kilowatt-hours, which enters Denmark's main power grid. Meanwhile, Samsø's petroleum-fueled transportation sector uses the equivalent of just 53 million kilowatt-hours per year, so the islanders figure they're still in the cosmic clear with room to spare.

The bottom line: Samsø is 140 percent carbon negative, while virtually the rest of the universe — except for off-the-grid pockets in a few communities and the solar-powered International Space Station — is carbon positive to the tune of adding 27 billion metric tons of CO_2 to the atmosphere each year. (Afghanistan and Chad are among the nations with the lowest per-capita carbon emissions; the United States releases the most overall — nearly one-quarter of the planet's total.)

Someday Samsø hopes to use its surplus power to make hydrogen or charge lithium-ion batteries to run its cars. In the meantime, I can thumb a teeny car ride (three miles) in the rain. It's offset! I eat a thick slice of beef at a smorgasbord. Even though the meat probably originated in Argentina, I am guilt-free. Samsø's clean power flowing into the Denmark grid means that some coal plant doesn't have to produce the wattage. Sorry, Bente. Please pass the roast.

Jørgen Tranberg is a gap-toothed dairy and pumpkin farmer. His latest crop is the one-megawatt turbine I climbed. When I first meet him, he's trying to U-turn a tractor carrying a huge bale of hay. "I'm not an environmentalist," he says, flicking ash off his cigarette. "It's a business venture." Down the driveway sits his new Drago Fiesta motorboat, so I'm not surprised when he tells me business is good.

Almost everyone on Samsø, in one way or another, makes money from wind. Turbines are owned by private investors like Tranberg, by the government, or by cooperatives of people who bought shares to finance their construction. The process is democratic in the way so many things are in Denmark; shares cost about $360 each.

Tranberg, for his part, took out a loan to buy his $1 million windmill six years ago, but the government guaranteed him an above-market price for his power. And the wind, which blows lustily here most days of the year, proved to be an even better friend than he and other islanders had hoped. Investors have seen returns of 8 percent or so a year, which works out to roughly $100,000 per on-shore turbine. Tranberg's is already paid off. "It's enough income for me that I don't have to work, but I like to work," he says. Besides, he adds, talking tough for a man in clogs, "we can't put all that shit in the sky from coal. There's too much shit in the air."

The fairy tale of turning pumpkins into turbines is one told frequently on the isle, most notably in the Ecomuseum in the main town of Tranebjerg. There, among the Viking coins and the potato-shaped chocolate bonbons in the gift shop, are posters that explain the history of the Samsø experiment: how in 1997 the green-leaning government and the Danish Energy Agency launched a contest to select the island with the best plan to become energy-sustainable by 2008; how Samsø won and has since spent $70 million on the project; and how European grants also helped subsidize the effort.

The result has been fairly staggering. Between 1997 and 2005, Samsø is estimated to have reduced emissions of global-warming pollutants like carbon dioxide, sulfur dioxide, and nitrous oxide by 142, 71, and 41 percent, respectively. Today 100 percent of the islanders use green electricity, and 70 percent of the residences are heated by wind, solar, or biomass systems, including huge, centrally located furnaces that burn straw. Straw is considered carbon neutral because it's a byproduct of crops like alfalfa that help absorb CO_2 from the atmosphere, just as trees do. The straw is burned in super-hot, super-efficient kilns. Plus the ash is collected and spread back on the fields to fertilize the next cycle. The cows eat the alfalfa, make some damn fine milk, and then the straw goes off to the furnace.

It's so Scandinavian.

I'm staying down the road a few miles with another dairy farmer, Erik Andersen, and his wife, Lise, in the tiny town of Besser. Now in their sixties, they live with their Border collie, Jacko, in a beautiful, simple stucco house. Andersen, who has milked cows all his life, caught the renewables bug in a big way after Samsø won the energy contest. He installed eighteen square yards of solar paneling on his barn roof, supplying all his electricity and most of his heat and hot water from April to October. The rest of the year he chops scrap wood from his land to feed his biomass furnace. He has also taken to growing rapeseed, otherwise known as canola, for home-brewed biodiesel.

After serving me aeblekage, a baked apple cake with otherworldly good fresh cream, Andersen shows off his voltage meter. "I like to come out and watch the meter running backward," he says. He is tall and lumbering and sports a gray buzz cut. Every day he wears overalls and irrigation boots, and he is not a man of many words. "It is interesting to make your own energy," he says.

Next he takes me out to the garage to see his prized new possession: a rapeseed press. The black, beady seeds funnel into the machine, which squeezes the bejeebers out of them. One spout kicks out a narrow, cylindrical green mash, the byproduct cake, while another spout pours a wan, yellowish oil. Andersen feeds the nutritious mash to his cows so he doesn't have to import soy-based feed. He then filters and pours the oil right into his two tractors and his red Volkswagen Passat. No refining, no cleaning, no mixing with lye or other chemicals.

"What are we going to do when the oil runs out? You need to prepare," he says. By not buying petrol, which costs $6.38 per gallon here, he figures he's satisfying his independent streak.

"Try it," he says, grinning. He wants me to eat his gasoline. Fine. I pour some onto my finger and take a taste. It's nice — light and bland, kind of like oil and salad at the same time.

I'm not the only energy tourist on Samsø. Some 2,000 turbine peepers visit the place every year. The island has become a training ground for experts from Thailand, China, Europe, and America, including a group from the University of Virginia. Demand is so great that the town government is helping build a conference center (out of sustainable materials and in the shape of a Viking longhouse) to accommodate the eco-wonks who want to glean the island's secrets.

The week I'm here, a delegation from Ireland and Scotland is also. We all watch a former vegetable farmer named Søren Hermansen give a PowerPoint presentation at an old seaside mansion, surrounded by the sound of lapping waves. After a brief demonstration of a Barbie-doll-scale toy hydrogen car — it works, sputteringly — we step outside. The lulling view is broken only by a passing Russian oil tanker bound for a place that is most decidedly not Samsø.

If this is the Fantasy Island of the greenie set, then Hermansen is its Ricardo Montalban. Instead of white suits, though, he wears denim, and it's not hard to picture him in one of those Norse helmets with horns. His ancestors probably wore them, because his family has lived on Samsø for at least ten generations. Part miracle worker, part gracious host, he directs the European Union–backed Samsø Energy Office and has been with the renewable-energy project since its beginning. A part-time rock musician — he plays bass in a band called, fittingly, the Generators — Hermansen has earned the public's trust over nearly a decade of meetings, tours, and ribbon-cutting ceremonies.

In addition to orchestrating the development of Samsø's centralized heating plants, which efficiently warm 1,000 residences, he convinced hundreds of retirees to put government-subsidized renewable-energy systems in their homes. In the meantime, among other projects, he's shepherding farmers into a rapeseed-growing cooperative and overseeing a grant from the EU to install a hydrogen-production test facility to power vehicles. Hermansen is determined to extend the green dream to transportation, the one dirty sector left. "The vision is that all the gas stations on the island would pump plant oil," he says.

After a day, the Irish delegates are awed. "These lads here are light-years ahead of us," says Eugene Houlihan, who works for a building-supplies cooperative in Inis Oírr. He shakes his fair head. "We don't have a clue. Cheesus. So we're here to learn how to apply it," he says. "We are ashamed to listen to the Danes. Cheesus."

Well, hey, I point out as we sit in a pub overlooking the picturesque harbor, at least Ireland signed the Kyoto agreement. America can't even accept a cap on greenhouse-gas emissions. "True," he says, brightening. But then Anthony McCarthy, an energy consultant from County Wexford, pipes in: "If Americans get serious about alternative energy, they will fucking pass everybody." He peers deep into his Carlsberg. "That is the beauty of America."

After they were done invading the civilizations of northern Europe, the industrious Danes turned to building great ships, first out of wood, then steel. Now they employ a similar metal-and-rivet fabrication method to make windmills for Vestas, one of the largest turbine manufacturers in the world. This, after all, is also the birthplace of Lego; Danes are organized, rational, and easily assembled. The highest earners pay about a 60 percent income tax, and what they get in return is free universities, free medical care, and some really good, cheap cheese.

But how did Danes get to be so carbon goody-goody? In Copenhagen, I ask the man who should know: Lego chairman Mads Øvlisen, who also serves on the United Nations' newly created Global Compact Board and drives an electric Citroën to work. After an eloquent briefing on the Vikings — "Danes believe they can change the world, and, therefore, they must!" — Øvlisen tells me that Denmark's ability to look outward stems from its seafaring past. That, and its tradition of social equity, made sustainability highly desirable. "We are a tribal country," he says. "And that drives innovation and adaptation."

Geopolitics also played a part. In 1973 OPEC imposed an oil embargo against the United States and the Netherlands for supporting Israel's war against Syria and Egypt and nearly quadrupled the price of petroleum for everyone else. The United States at the time imported about 35 percent of its oil; Denmark imported more than 90 percent. Keenly aware of their vulnerability, the Danes spent the next thirty years figuring out how to secure an energy-independent future, all without nuclear power, which parliament outlawed in 1982.

Now, remarkably, Denmark is about 150 percent self-sufficient. A net exporter of energy, most of it oil and natural gas from the North Sea, it also sells wind power to it neighbors.

The American response to the embargo, by comparison, was more of a cheap-oil-is-our-birthright hissy fit. In the late 1970s and early '80s, to be fair, the United States launched some serious alternative-energy schemes, complete with tax credits and federal funding for renewables, and for a brief moment California actually became the king of wind power. Then, during the Reagan era, federal and state subsidies expired, making it impossible for wind and solar to compete with oil and coal. Today the United States imports nearly double the oil that it did in 1973 — or about 60 percent of what it consumes. Wind power makes up less than 1 percent of the American electricity pie.

We can turn this around if we choose to, observers note. "This country is capable of sleeping for a long time," says the former CIA chief James Woolsey, now an eco-minded energy-security specialist and vice president of Booz Allen Hamilton, a Virginia-based international consulting firm. "But when it wakes up, things move pretty fast, and that's what we're going to be seeing in the next few years, with a growing coalition of tree huggers, do-gooders, cheap hawks, evangelicals, venture capitalists, and Willie Nelson. Renewable-energy technologies," Woolsey adds, "can come faster than people realize."

Indeed, in western Indiana, of all places, there's a mini-Samsø rising from the cornfields: Reynolds, population 550, is constructing a $9.5 million power plant to turn hog manure and other agricultural products into clean energy. Elsewhere, at least 333 mayors representing more than 54 million Americans have pledged to follow the Kyoto mandates for reducing greenhouse gases, without waiting for Washington.

But renewables like wind power still face some Hummer-size hurdles. "In the United States," says environmentalist Frances Beinecke, president of the Natural Resources Defense Council, "people still see wind power as futuristic and marginal." To that list one could also add "ugly": a proposed wind farm off Nantucket, for example, faces strong opposition from locals who don't want turbines marring the view.

"Today we can point to Denmark and say, 'Hey, a Western country gets one-fifth of its electricity from wind.' It's doable," Beinecke says. "There's just a cultural divide we have to cross."

On my last day on Samsø, I ride my bike around the island. With its waving grasses and gentle surf, I can see why Valdemar the Victorious gave the place to his new queen as a postnuptial gift in 1214. Today goats are grazing contentedly under the photovoltaic array at a solar heating plant. I walk through some sunflowers to visit farmers who raise organic pigs and perfect cherry tomatoes. Leaving for the mainland, the ferry passes a chain of windmills, their graceful arms performing an unsynchronized water ballet.

All too soon, I'm back in carbon-positive Copenhagen. My own CO_2 balance sheet is, regrettably, back in play. I'm going to be more than two carbon tons in the red, according to eco-entrepreneur Shea Gunther, cofounder of Renewable Choice Energy, a Boulder,

Colorado–based firm that for about $65 will buy U.S. wind power to offset my trip for me. (The final tally: my 10,740-mile round-trip flight from Seattle to Copenhagen, plus airport commutes, created 4,315 pounds of CO_2; my train trips made 50 pounds; and my laptop and cell phone set me back 9. And that's not even including the mojito.) So for my last night in Denmark, I really should return to the infernally correct Don't-Even-Think-About-Catching-Sleep-in Green.

But I just can't bring myself to face sixty-six other eyeballs in my room. My days in Samsø were too full of guilt-free delights and what the Danish call *hygge,* or coziness. I get to the city after dark. It's late and I'm hungry. I start bicycling toward the hostel but pass a charmingly weird lodge called the Hotel Kong Arthur. Its flagrantly inefficient incandescent light bulbs emit a friendly, warm glow. There's a suit of armor by the entrance and soft upholstery decorated with fleurs-de-lis. How can I resist? I can't. It includes a free breakfast! And a private bathroom!

I rant to myself that this is exactly why governments should step in and support responsible energy development: so that wayward, flawed sybarites such as myself don't have to make endless, irksome daily decisions when all we really want is a warm bed and a bowl of muesli. With a side of bacon. And a mug of hot imported coffee. *Ja.*

Contributors' Notes

Jon Cohen is a correspondent with *Science* and has also written for *Outside, The New Yorker,* the *Atlantic Monthly,* the *New York Times Magazine, Smithsonian, Slate, Technology Review, Discover,* the *Washington Post, New Republic, Glamour, Surfer,* and other publications. He has published two books, *Shots in the Dark: The Wayward Search for an AIDS Vaccine* and *Coming to Term: Uncovering the Truth about Miscarriage.* He is currently working on a book that will examine how new research is redrawing the dividing lines between humans and chimpanzees.

John Colapinto is a staff writer at *The New Yorker* and the author of the nonfiction book *As Nature Made Him: The Boy Who Was Raised as a Girl* and the novel *About the Author.* He is married and has a nine-year-old son.

Christopher J. Conselice is an astronomer who is currently an associate professor and reader in the School of Physics at the University of Nottingham in the UK. He is an expert in the field of galaxy formation, studying the earliest galaxies and the relation of galaxies to the cosmological features of the universe. He did his undergraduate, graduate, and postdoctoral work at the University of Chicago, the University of Wisconsin, and the California Institute of Technology, respectively. He is a native of Jacksonville, Florida.

Gareth Cook is a Pulitzer Prize–winning journalist and the editor of the Ideas section of the *Boston Globe.* He graduated from Brown University with degrees in international relations and mathematical physics. He has worked at *Foreign Policy, U.S. News & World Report, Washington Monthly,* and the *Boston Phoenix.* In addition to the Pulitzer, he has won a National Academies Communication Award, a National Headliner Award, and an Ocean

Science Journalism Award from the Woods Hole Oceanographic Institution. He lives in Jamaica Plain, Massachusetts, with his wife, Amanda, and their two sons, Aidan and Oliver.

C. Josh Donlan is a visiting fellow at Cornell University and the director of Advanced Conservation Strategies, an NGO dedicated to reversing biodiversity loss through the development of innovative, self-sustaining, and economically efficient solutions to environmental challenges. Josh works on a variety of issues, including understanding how species interact and what consequences those interactions have on the ecosystem; reversing the impact of invasive species; incorporating ecological history into conservation strategies; and developing financial and incentive instruments for conservation. Josh spends much of his time restoring islands around the world and has worked in Argentina, Australia, Ecuador, Chile, Mexico, and the United States. He served for four years as the science and conservation adviser to the Galapagos National Park, where he helped run the world's largest island restoration project.

Freeman Dyson is a retired professor at the Institute for Advanced Study in Princeton. Born in England in 1923, he moved to the United States in 1951. He spent the first half of his life mostly doing research in mathematics and physics, and the second half mostly writing books for the general public. His most recent books are *The Scientist as Rebel* and *A Many-Colored Glass,* from which "Our Biotech Future" was excerpted.

Steve Featherstone is a writer and photographer. He lives in upstate New York.

Michael Finkel lives with his family in western Montana. He has been on assignment in more than forty nations and is curious about pretty much everything.

James Geary is a regular contributor to *Popular Science* and the author of the *New York Times* bestseller *The World in a Phrase: A Brief History of the Aphorism* as well as *Geary's Guide to the World's Great Aphorists* and *The Body Electric: An Anatomy of the New Bionic Senses.*

Robin Marantz Henig has been a freelance writer for twenty-five years and has written eight books. Her most recent, *Pandora's Baby,* is about the early days of in vitro fertilization research and the controversy surrounding the creation of the world's first test-tube baby.

Edward Hoagland is seventy-five and has published twenty books. He began writing professionally in 1951 when he joined the Ringling Brothers and Barnum & Bailey Circus and wrote his first novel about that experience. He has traveled to Africa five times and to Alaska and British Columbia nine times, as well as to Yemen, India, Antarctica, and other countries. "Life is an ecstasy," as Emerson said.

Olivia Judson is an evolutionary biologist at Imperial College London. Under the pseudonym Dr. Tatiana, she wrote a best-selling guidebook to sex throughout the natural kingdom called *Dr. Tatiana's Sex Advice to All Creation*.

Walter Kirn is an American novelist and critic who lives in Montana. A 1983 graduate of Princeton University, he has published a collection of short stories and several novels, including *Thumbsucker,* which was made into a 2005 film featuring Keanu Reeves and Vince Vaughn; *Up in the Air;* and *Mission to America.*

Andrew Lawler is a senior writer with *Science* and a freelance writer for *Smithsonian, National Geographic, Discover, Archaeology, Audubon, American Archaeology, Air & Space,* the *Columbia Journalism Review,* and other magazines. He has written hundreds of articles on topics ranging from asteroids to zebra fish and has filed stories from many countries, from Afghanistan to Yemen. His work has appeared twice in *The Best American Science and Nature Writing,* and a third article received an honorable mention. When not waiting for his luggage at airport carousels, he lives in rural Maine.

Jon Mooallem has written for the *New York Times Magazine, Harper's Magazine, Salon, Mother Jones,* and other publications. He lives in San Francisco.

Ian Parker is a British journalist who lives in New York. He has been a staff writer at *The New Yorker* since 2000.

Todd Pitock's work has appeared in many publications, including the *Washington Post, Discover, Salon,* the *New York Times,* and *ForbesLife,* where he is a contributing editor and columnist. A winner of numerous awards, including, most recently, the American Society of Journalists and Authors Award for reporting on a significant topic and the Lowell Thomas Award for travel writing, he has spent five years in Israel and South Africa and has reported from Iraq, Libya, Antarctica, and dozens of other countries. He lives near Philadelphia with his wife, Toni, and their three children. His Web site is www.toddpitock.com.

David Quammen is the author of eleven books, including *The Song of the Dodo* and *The Reluctant Mr. Darwin.* He is a contributing writer for *National Geographic* and currently holds the Wallace Stegner Chair in Western American Studies at Montana State University. He has received the National Magazine Award three times, most recently for "Was Darwin Wrong?" in the November 2004 issue of *National Geographic*. He is at work on a book about zoonotic diseases.

Ron Rosenbaum has written for many publications, including *Esquire, Harper's Magazine, High Times, Vanity Fair,* and the *New York Times Magazine.* He spent more than ten years doing research on Adolf Hitler, which included traveling to Vienna, Munich, London, Paris, London, and Jerusalem and interviewing leading historians, philosophers, biographers, theologians, and psychologists. The result was his acclaimed 1998 book, *Explaining Hitler: The Search for the Origins of His Evil.* His latest book is *The Shakespeare Wars.*

Oliver Sacks is a physician and the author of nine books, including two collections of case histories, *The Man Who Mistook His Wife for a Hat* and *An Anthropologist on Mars,* in which he described patients struggling to adapt to various neurological conditions. His book *Awakenings* inspired the Oscar-nominated film of the same name and the play *A Kind of Alaska,* by Harold Pinter. He practices neurology in New York City.

Michael Specter, who has been on the staff of *The New Yorker* since 1998, writes frequently about science, public health, and the impact of new technologies on society.

Jeffrey Toobin is a legal analyst for CNN and a staff writer at *The New Yorker.*

Andreas von Bubnoff is a science journalist based in New York City. He has written for the *Chicago Tribune,* the *Los Angeles Times,* the Swiss newspaper *SonntagsZeitung, Prevention, Science News, Nature, Cell,* and other publications. He earned a Ph.D. in developmental biology, spending his days (and many nights) in a research lab making glow-in-the-dark transgenic frogs.

Florence Williams is a contributing editor at *Outside* and frequently writes for the *New York Times* and the *New York Times Magazine.* She holds a B.A. from Yale and a degree in creative writing from the University of Montana. A three-time winner of awards from the American Society of Journalists and Authors, she was a Ted Scripps Fellow at the University of Colorado's Center for Environmental Journalism in 2007–2008.

Other Notable Science and Nature Writing of 2007

SELECTED BY TIM FOLGER

IVAN AMATO
Experiments of Concern. *Chemical & Engineering News,* July 30.

REMI BENALI AND GEORGE BLACK
The Real Price of Gold. *OnEarth,* Spring.

BURKHARD BILGER
Spider Woman. *The New Yorker,* March 5.

KENNETH BROWER
When Ants Can Fly. *California Magazine,* September/October.

CLIFF BURGESS AND FERNANDO QUEVEDO
The Great Cosmic Roller-Coaster Ride. *Scientific American,* November.

AKIKO BUSCH
Swimming the Hudson. *Harvard Review,* Spring.

PETER CANBY
Deadly Sonar. *OnEarth,* Spring.

CHRIS CARROLL
End of the Line. *National Geographic,* April.

RON COWEN
Bang. *National Geographic,* March.

GREG CRITSER
Of Men and Mice. *Harper's Magazine,* December.

ANDREW CURRY
The Dawn of Art. *Archaeology,* September/October.

ANTHONY DOERR
Window of Possibility. *Orion,* July/August.

ROSS DOUTHAT
The God of Small Things. *The Atlantic,* January/February.

FREEMAN DYSON
Freedom of Inquiry. *Bulletin of the Atomic Scientists*, January/February.

JOSHUA FOER
Remember This. *National Geographic*, November.
ABIGAIL FOERSTNER
What Van Allen Found in Space. *Bulletin of the Atomic Scientists*, July/August.
DOUGLAS FOX
Consciousness in a . . . Cockroach? *Discover*, January.
MARGALIT FOX
Village of the Deaf. *Discover*, July.
PETER FRIEDERICI
Facing the Yuck Factor. *High Country News*, September 17.
MCKENZIE FUNK
Cold Rush. *Harper's Magazine*, September.

ATUL GAWANDE
The Way We Age Now. *The New Yorker*, April 30.
JOSIE GLAUSIUSZ
Toxic Salad. *Discover*, April.
GARY GREENBERG
Manufacturing Depression. *Harper's Magazine*, May.

BERND HEINRICH AND THOMAS BUGNYAR
Just How Smart Are Ravens? *Scientific American*, April.
DONOVAN HOHN
Moby-Duck. *Harper's Magazine*, January.

BARBARA KINGSOLVER
Stalking the Vegetannual. *Orion*, March/April.
ELIZABETH KOLBERT
Crash Course. *The New Yorker*, May 14.
KEVIN KRAJICK
Small Miracles of the Cave World. *OnEarth*, Summer.

ROBERT LANZA
A New Theory of the Universe. *The American Scholar*, Spring.
SETH LLOYD
You Know Too Much. *Discover*, April.

PETER MAASS
Slick. *Outside*, March.
LARISSA MACFARQUHAR
Two Heads. *The New Yorker*, February 12.
SUSAN MCGRATH
Corkscrewed. *Audubon*, January/February.

THE BEST AMERICAN SERIES®

THE BEST AMERICAN SHORT STORIES® 2008
Salman Rushdie, editor, Heidi Pitlor, series editor
ISBN: 978-0-618-78876-7 $28.00 CL
ISBN: 978-0-618-78877-4 $14.00 PA

THE BEST AMERICAN NONREQUIRED READING™ 2008
Edited by Dave Eggers, introduction by Judy Blume
ISBN: 978-0-618-90282-8 $28.00 CL
ISBN: 978-0-618-90283-5 $14.00 PA

THE BEST AMERICAN COMICS™ 2008
Lynda Barry, editor, Jessica Abel and Matt Madden, series editors
ISBN: 978-0-618-98976-8 $22.00 POB

THE BEST AMERICAN ESSAYS® 2008
Adam Gopnik, editor, Robert Atwan, series editor
ISBN: 978-0-618-98331-5 $28.00 CL
ISBN: 978-0-618-98322-3 $14.00 PA

THE BEST AMERICAN MYSTERY STORIES™ 2008
George Pelecanos, editor, Otto Penzler, series editor
ISBN: 978-0-618-81266-0 $28.00 CL
ISBN: 978-0-618-81267-7 $14.00 PA

THE BEST AMERICAN SPORTS WRITING™ 2008
William Nack, editor, Glenn Stout, series editor
ISBN: 978-0-618-75117-4 $28.00 CL
ISBN: 978-0-618-75118-1 $14.00 PA

THE BEST AMERICAN TRAVEL WRITING™ 2008
Anthony Bourdain, editor, Jason Wilson, series editor
ISBN: 978-0-618-85863-7 $28.00 CL
ISBN: 978-0-618-85864-4 $14.00 PA

THE BEST AMERICAN SCIENCE AND NATURE WRITING™ 2008
Jerome Groopman, editor, Tim Folger, series editor
ISBN: 978-0-618-83446-4 $28.00 CL
ISBN: 978-0-618-83447-1 $14.00 PA

THE BEST AMERICAN SPIRITUAL WRITING™ 2008
Edited by Philip Zaleski, introduction by Jimmy Carter
ISBN: 978-0-618-83374-0 $28.00 CL
ISBN: 978-0-618-83375-7 $14.00 PA